ENERGY SCIENCE, ENGINEERING AND TECHNOLOGY

BIOFUELS

ADVANCES IN RESEARCH AND APPLICATIONS

Energy Science, Engineering and Technology

Additional books and e-books in this series can be found on Nova's website under the Series tab.

ENERGY SCIENCE, ENGINEERING AND TECHNOLOGY

BIOFUELS

ADVANCES IN RESEARCH AND APPLICATIONS

GEORGE R. CAREY
EDITOR

Copyright © 2020 by Nova Science Publishers, Inc.

All rights reserved. No part of this book may be reproduced, stored in a retrieval system or transmitted in any form or by any means: electronic, electrostatic, magnetic, tape, mechanical photocopying, recording or otherwise without the written permission of the Publisher.

We have partnered with Copyright Clearance Center to make it easy for you to obtain permissions to reuse content from this publication. Simply navigate to this publication's page on Nova's website and locate the "Get Permission" button below the title description. This button is linked directly to the title's permission page on copyright.com. Alternatively, you can visit copyright.com and search by title, ISBN, or ISSN.

For further questions about using the service on copyright.com, please contact:
Copyright Clearance Center
Phone: +1-(978) 750-8400   Fax: +1-(978) 750-4470   E-mail: info@copyright.com.

NOTICE TO THE READER

The Publisher has taken reasonable care in the preparation of this book, but makes no expressed or implied warranty of any kind and assumes no responsibility for any errors or omissions. No liability is assumed for incidental or consequential damages in connection with or arising out of information contained in this book. The Publisher shall not be liable for any special, consequential, or exemplary damages resulting, in whole or in part, from the readers' use of, or reliance upon, this material. Any parts of this book based on government reports are so indicated and copyright is claimed for those parts to the extent applicable to compilations of such works.

Independent verification should be sought for any data, advice or recommendations contained in this book. In addition, no responsibility is assumed by the Publisher for any injury and/or damage to persons or property arising from any methods, products, instructions, ideas or otherwise contained in this publication.

This publication is designed to provide accurate and authoritative information with regard to the subject matter covered herein. It is sold with the clear understanding that the Publisher is not engaged in rendering legal or any other professional services. If legal or any other expert assistance is required, the services of a competent person should be sought. FROM A DECLARATION OF PARTICIPANTS JOINTLY ADOPTED BY A COMMITTEE OF THE AMERICAN BAR ASSOCIATION AND A COMMITTEE OF PUBLISHERS.

Additional color graphics may be available in the e-book version of this book.

Library of Congress Cataloging-in-Publication Data

ISBN: 978-1-53617-721-3

Published by Nova Science Publishers, Inc. † New York

CONTENTS

Preface vii

Chapter 1 Non-Catalytic Production of Biodiesel:
Energy and Safety Considerations 1
*Fernando Israel Gómez-Castro
and Antioco López-Molina*

Chapter 2 Thermodynamic Properties of Biofuels:
Comparison and Review of Excess Enthalpy
of Mixtures of Butanol, or Dibutylether,
with Representative Hydrocarbons 31
*F. Aguilar, N. Muñoz-Rujas, E. Montero
and F. E. M. Alaoui*

Chapter 3 Environmental Aspects of Using Biodiesel as a
Sustainable Energy Source:
Current Situation and Future Trends 57
*Mehdi Ardjmand, Farid Jafarihaghighi,
Mehrdad Mirzajanzadeh, Aida Gifani
and Hasanali Bahrami*

Chapter 4	Non-Conventional Yeasts with Potential for Production of Second-Generation Ethanol *Katharina O. Barros, Angela M. Garcia-Acero and Carlos A. Rosa*	**109**
Index		**151**

PREFACE

Biofuels: Advances in Research and Applications first explores previously studied supercritical processes for the production of biodiesel. Two of the main drawbacks of said processes are discussed: the high energy requirements and the hazards due to the conditions under which the processes are operated.

The authors present a review of the experimental excess enthalpy of mixtures of dibutyl ether or butanol, with representative hydrocarbons. The most relevant functional groups of gasoline hydrocarbon types are considered: heptane, iso-octane, 1-hexene, cyclohexane, methylcyclohexane, benzene, and toluene.

Continuing, this compilation briefly explores the production of biofuel from different edible and non-edible feedstock, mentioning the various types of homogeneous and heterogeneous acid or base catalysts applied for reactions. The benefits and drawbacks of biodiesel as compared to diesel fuel are also described.

Several yeasts are examined for their capability to produce xylanolytic enzymes that degrade xylan, the major polysaccharide in the hemicellulose structure. The production of hydrolytic enzymes for the enzymatic hydrolysis process is reported by optimizing lignocellulose degradation and increasing the yield of simple sugars.

Chapter 1 - The production of biofuels is an important topic in the new century, due to the need for satisfying the energy demand of the transport sector. One of the better-known biofuels is biodiesel, which is a mixture of alkyl esters, obtained by the chemical treatment of the triglycerides and fatty acids contained in vegetable oils and animal fats. Biodiesel is usually employed to run diesel engines, blended with fossil diesel. The traditional production of biodiesel involves the homogeneous transesterification of the triglycerides with an alcohol and an alkaline catalyst, usually sodium or potassium hydroxide. Nevertheless, if the raw material has a relatively high content of free fatty acids, the alkaline catalyst may cause undesired reactions, as saponification; thus, a pre-treatment is necessary to transform the fatty acids into alkyl esters with an acid catalyst, commonly sulfuric acid. This implies the need for additional purification steps to neutralize the stream leaving the reactor. To overcome such situations, the use of supercritical alcohols to produce biodiesel has been proposed. By this supercritical approach, the formation of two phases in the reactor is avoided, enhancing the mass transfer and reducing the required residence time. Moreover, since there is no need for a catalyst, the purification step necessary to obtain the pure biodiesel is expected to have less equipment. The process with a single reactor using supercritical methanol as reactant is one of the most known alternatives, although there are other reported approaches, as the hydrolysis-esterification process, by which lower temperatures are required, and other proposals using non-alcoholic supercritical reactants. In this chapter, some previously studied supercritical processes to produce biodiesel are presented and described. Two of the main opportunity areas for such processes are discussed: the high energy requirements and the hazards due to the conditions under which the processes are operated. The use of tools as process intensification and process integration is presented as a strategy to reduce energy requirements in the processes. Moreover, an analysis of the main risks and some suggestion for inherent safety design in the supercritical processes are presented. Finally, the use of computational tools is presented as a way to reduce the inherent risk in such processes, through the determination of safety policies from the design stage.

Chapter 2 - The use of alcohols and ethers as oxygenated compounds in gasoline blends has been proposed in order to reduce the emmisions of new reformulated gasoline. Alcohol, or ether + hydrocarbon mixtures are of interest as model mixtures for gasoline in which the alcohol and the ether act as non-polluting, high octane number blending agents. Higher alcohols (those containing more than two carbon atoms) coming from renewable sources, and several other oxygenated compounds, can be used as blend components in gasoline for the reduction of petroleum consumption and greenhouse gas emissions. Amongst these oxygenates, butanol is considered as an alternative to conventional gasoline and diesel fuels. Butanol has many advantages over other potential alternative fuel candidates such as ethanol. Butanol can be demonstrated to work in the internal combustion engines designed for use with gasoline without modification at a composition rate of 85% in volume (unlike 85% ethanol, E85). Butanol presents a similar contribution to the antiknock effect to those of methanol and ethanol, while its energy content per unit volume is higher than ethanol, and almost as high as gasoline. Butanol is less susceptible to separation in the presence of water than ethanol/gasoline blends, therefore allowing the use of the existing distribution infrastructure without requiring modifications to blending facilities, storage tanks or retail station pumps. 1-Butoxybutane, also known as dibutyl ether (DBE) is considered to be a valuable additive to second generation bio-fuels. It acts as a non-polluting, high octane number compound used as a blending agent in reformulated gasoline. And butanol is a basic component in the synthesis of the ether, and therefore is always present as an impurity. Thus, the mixtures of DBE, or butanol, with the hydrocarbons representative of the gasolines is a topic of interest in the modelling and prediction of properties of future biofuels. Moreover, for the aim of optimizing common industrial processes (storage, transport, separation and mixing processes), reliable experimental data are needed. Accurate empirical equations, models and simulation programs need to be fed with such experimental data to be useful. This work presents a review of the experimental excess enthalpy (or enthalpy of mixing) of mixtures of DBE, or butanol, with representative hydrocarbons. The most relevant functional groups of gasoline hydrocarbon types are considered: heptane (alkane), iso-octane (branched alkane), 1-hexene (alkene), cyclohexane (cyclic), methylcyclohexane (branched cyclic), benzene (aromatic), and toluene (branched aromatic). Excess enthalpy of mixtures is a valuable property when evaluating the mixing and storage behaviour of fuels. The excess enthalpies of a set of 14 binary systems of oxygenated + hydrocarbon experimentally determined by the authors are presented and compared, at 298.15 and 313.15 K. In addition, a literature review on the excess enthalpy of mixtures of 1-butanol, or dibutyl ether, with representative hydrocarbons is included. This review could be of interest for the biofuel industry, within the production, transport and end-user (automotive) sectors.

Chapter 3 - The current energy crisis in the era of increasing energy consumption, together with the increment in greenhouse gas concentrations (for instance: carbon dioxide, methane, and nitrous oxide, etc.) from burning petroleum-based fuels causing environmental issues, have made scientist consider substituting fossil fuels with renewable and clean fuels. Hence, great studies have been conducted regarding finding alternative fuels. Amongst fuels like methanol, liquefied petroleum gas (LPG), compressed natural gas (CNG), vegetable oils, liquefied natural gas (LNG), reformulated diesel fuel, and reformulated gasoline, vegetable oils are the only non–fossil fuels. Biodiesel is one of the promising alternatives derived from renewable resources, like animal fats or vegetable oils, which have been realized as an environmentally benign fuel because of not only its green advantages (sustainability, biodegradability, and non-toxicity), but also several extra social benefits, like creation of new jobs, less emission of soot and carbon, and less global warming. In this regard, an attempt has been made to briefly introduce the production of this fuel from different edible and non-edible feedstock to mention various types of homogeneous and heterogeneous acid or base catalyst applied for reactions. The benefits and drawbacks of biodiesel compared with diesel fuel are also included. Some of the advantages of biodiesel are 1-carbon cycle, 2- less emission, 3- improvement of combustion and reduction of emissions of unburned hydrocarbons due to oxygen presence, 4- higher lubricity, 5- higher flash point and higher cetane number, and etc.

Chapter 4 - Second-generation (2G) ethanol production is dependent on the efficient conversion of the carbohydrates from the cellulosic and hemicellulosic fractions of lignocellulose, primarily D-glucose and D-xylose. Although, pentose is present in high levels in this substrate, *Saccharomyces cerevisiae*, the most widely used microorganism for industrial alcoholic fermentation, cannot metabolize this sugar. Non-conventional yeast species that are able to ferment xylose have been found in habitats such as rotting wood, tree bark, and leaves, and some of them are even associated with insects. *Pachysolen tannophilus* was the first species described to convert xylose directly to ethanol. Thereafter, other xylose-fermenting yeasts able to ferment this pentose to ethanol were isolated and identified. Among these, the ones that stand out are the species of the clades *Scheffersomyces* and *Spathaspora* as *Sc. stipitis*, *Sc. shehatae*, *Sc. lignosus*, *Sp. passalidarum*, *Sp. arborariae*, *Sp. gorwiae*, *Sp. hagerdaliae*, and *Sp. piracicabensis*. *Spathaspora passalidarum* was isolated from the gut of host beetles and rotting wood, and it is considered the most prominent species for the transformation of xylose to ethanol. Several yeasts are also capable of producing xylanolytic enzymes that degrade xylan, the major polysaccharide in the hemicellulose structure. The production of hydrolytic enzymes for the enzymatic hydrolysis process is reported by optimizing lignocellulose degradation and increasing the yield of simple sugars. Some basidiomycetous and ascomycetous yeasts produce xylanases and β-xylosidases from substrates such as xylan and D-xylose. The yeast-like fungus *Aurobasidium pullulans* is known to be a source of xylanolytic enzymes with high specificity. It can be used for simultaneous saccharification and fermentation with xylose-fermenting yeasts to improve the production of lignocellulosic ethanol. Xylanolytic yeasts also include species of the clades *Sugiyamaella* (*Su. xylanicola, Su. lignohabitans, Su. valenteae*, and *Su. smithiae*), *Lodderomyces/Candida albicans* (*C. tropicalis*), and *Clavispora/Candida* (*C. intermedia*); and the genera *Scheffersomyces* (*Sc. shehatae* and *Sc. sipitis*), *Naganishia* (*N. diffluens*), *Kwoniella* (*Kw. heveanensis*), and *Papiliotrema* (*Pa. laurentii*). Yeasts with the ability to ferment sugars of lignocellulose and/or produce enzymes that act on this substrate have great potential for being applied to the production of 2G ethanol.

In: Biofuels
Editor: George R. Carey

ISBN: 978-1-53617-721-3
© 2020 Nova Science Publishers, Inc.

Chapter 1

NON-CATALYTIC PRODUCTION OF BIODIESEL: ENERGY AND SAFETY CONSIDERATIONS

Fernando Israel Gómez-Castro[1,*]
and Antioco López-Molina[2]

[1]Departmento de Ingeniería Química, Universidad de Guanajuato, Guanajuato, Guanajuato, Mexico
[2]División Académica Multidisciplinaria de Jalpa de Méndez, Universidad Juárez Autónoma de Tabasco, Jalpa de Méndez, Tabasco, Mexico

ABSTRACT

The production of biofuels is an important topic in the new century, due to the need for satisfying the energy demand of the transport sector. One of the better-known biofuels is biodiesel, which is a mixture of alkyl esters, obtained by the chemical treatment of the triglycerides and fatty acids contained in vegetable oils and animal fats. Biodiesel is usually employed to run diesel engines, blended with fossil diesel. The traditional production of biodiesel involves the homogeneous transesterification of

---

[*] Corresponding Author's E-mail: fgomez@ugto.mx.

the triglycerides with an alcohol and an alkaline catalyst, usually sodium or potassium hydroxide. Nevertheless, if the raw material has a relatively high content of free fatty acids, the alkaline catalyst may cause undesired reactions, as saponification; thus, a pre-treatment is necessary to transform the fatty acids into alkyl esters with an acid catalyst, commonly sulfuric acid. This implies the need for additional purification steps to neutralize the stream leaving the reactor. To overcome such situations, the use of supercritical alcohols to produce biodiesel has been proposed. By this supercritical approach, the formation of two phases in the reactor is avoided, enhancing the mass transfer and reducing the required residence time. Moreover, since there is no need for a catalyst, the purification step necessary to obtain the pure biodiesel is expected to have less equipment. The process with a single reactor using supercritical methanol as reactant is one of the most known alternatives, although there are other reported approaches, as the hydrolysis-esterification process, by which lower temperatures are required, and other proposals using non-alcoholic supercritical reactants. In this chapter, some previously studied supercritical processes to produce biodiesel are presented and described. Two of the main opportunity areas for such processes are discussed: the high energy requirements and the hazards due to the conditions under which the processes are operated. The use of tools as process intensification and process integration is presented as a strategy to reduce energy requirements in the processes. Moreover, an analysis of the main risks and some suggestion for inherent safety design in the supercritical processes are presented. Finally, the use of computational tools is presented as a way to reduce the inherent risk in such processes, through the determination of safety policies from the design stage.

Keywords: biodiesel, non-catalytic production, inherent safety

INTRODUCTION

One of the main concerns of mankind for several years has been to ensure the fulfillment of its energy requirements. Petroleum has been one of the most used energy sources since the end of the 1950s when the petroleum industry was born [1]. Nevertheless, several authors predict that the production of petroleum is diminishing [2]. Particularly, British Petroleum has reported that global petroleum consumption grew 1.7 million barrels per day in 2017, while its production was increased by only

0.6 million barrels per day [3]. On the other hand, there is a growing concern about the environmental impact derived from the continuous use of fossil fuels to produce energy. Burning fossil fuels releases several gases which are accumulated in the atmosphere, such as carbon dioxide, carbon monoxide, sulphur oxides, nitrogen oxides and methane, together with particles PM2.5 and PM10. From the Industrial Revolution, the concentration of such gases in the atmosphere has increased, accelerating the phenomena of global warming. According to the Environmental Protection Agency, global emissions of CO_2 into the atmosphere increased around 40% in the period 1990-2010 [4]; while British Petroleum reported that from 2016 to 2017, carbon emissions increased by 1.6% [3]. These two particular issues, the uncertainty on the petroleum availability and the concerns on the environmental impact due to the use of fossil fuels, caused a renewed interest on the development of renewable energy sources. The use of solar power, wind power, and other ways to produce renewable energy has considerably grown on recent years. In the case of the transport sector, one of the alternatives to turn into clean energies is through the use of liquid biofuels. Liquid biofuels are components which are produced from a renewable source and are expected to partially substitute fossil fuels. In the case of diesel engines, one of the candidates to be used as fuel is biodiesel, which is a mixture of alkyl esters obtained from sources with a high content of triglycerides, as vegetable oils or animal fats. Due to its chemical composition, it must be used in diesel fuels mixed with fossil diesel, with recommended compositions of biodiesel around 20 vol%, since a higher proportion of biodiesel may cause issues in cold climates.

The most common way to produce biodiesel is through basic catalysis, usually with sodium hydroxide; and an alcohol, usually methanol. The triglycerides in the raw material react with the alcohol producing alkyl esters and glycerol. One of the main advantages of this approach is that the reaction takes place at low temperatures, below 65°C [5]. Nevertheless, if the raw material has a high composition of free fatty acids, undesired saponification may occur, difficulting the purification of the biodiesel. In this case, an alternative is to perform a pre-treatment with sulfuric acid, esterifying the free fatty acids, then proceeding with the basic transesterification in a second reactor [6]. Nevertheless, acid pre-treatment implies additional neutralization requirements. This issue is particularly important when processing low-quality vegetable oils, e.g., waste cooking oils, since such kind of raw material represents an opportunity to reduce biodiesel production costs by 60-80% [7]; but those oils have a high composition of free fatty acids. To overcome such problems, the production of biodiesel with reactants under supercritical conditions has been proposed and studied. In supercritical processes, transesterification and esterification reactions may take place in the same device, the reaction rates are high, thus requiring less reactor volume, and the purification process is usually simplest than in the heterogeneously catalyzed processes [8]. On the other hand, the temperature required to perform the involved reactions is high, which implies high energy requirements to achieve such conditions. Moreover, another important issue which has not been studied in detail is the risk associated with the operational conditions. Recently, process safety has taken importance as an aspect to be included from the design stage. Some methodologies have been developed to assess the safety in a process, as the inherent safety approach [9]. In this chapter, a description of previously reported supercritical approaches for the production of biodiesel is presented. Some of the proposals to enhance the energetic performance of the supercritical processes are also described. Finally, an analysis of the main risks occurring in the supercritical processes for the production of biodiesel is developed, with a description of the methodologies that can be applied to the study of the supercritical systems and the proposal of strategies to enhance the safety in such processes.

NON-CATALYTIC PROCESSES FOR BIODIESEL PRODUCTION

Several non-catalytic reaction schemes have been proposed in recent years, and some of theme have been taken to a pilot or industrial scale, either physically or in a virtual environment. One of the first proposals for the production of biodiesel with methanol under supercritical conditions was presented by Kusdiana and Saka [10] for the transesterification of rapeseed oil, reporting that the reaction occurs at high rates for conditions of at least 350°C and 19 MPa. The first proposal for the complete process for the supercritical production of biodiesel was presented by Saka and Kusdiana [11]. Moreover, Kusdiana and Saka [12] proved that the water contains in the feedstock has little effect on the transesterification yield. Most raw materials for the production of biodiesel contain not only triglycerides but also free fatty acids. Thus, Warabi et al. [13] studied the characteristics of the transesterification of triglycerides and esterification of free fatty acids with several alcohols under supercritical conditions. Also, some works have proposed and analyzed industrial scale processes for the production of biodiesel under such technology in a simulation environment [14, 15]. Gómez-Castro et al. [15] concluded that the reaction conditions could be slightly reduced by changing the reactor design. A diagram of a production process for the production of biodiesel with supercritical methanol is shown in Figure 1. It can be seen that the process is quite simple and requires few equipments.

The use of ethanol as alternative reactant has also been proposed since it can be produced from biomass, while methanol is obtained from fossil sources. Gui et al. [16] reported as better conditions for the transesterification of ethanol 243°C and pressure over 6.38 MPa. Industrial-scale flowsheets have been proposed for such approach [17].

To reduce the pressure required to transform the vegetable oils into biodiesel, Saka proposed a two-step scheme, where sub-critical hydrolysis first occurs, transforming the triglycerides into fatty acids, then esterifying the free fatty acids with supercritical methanol [18]. For such process, required pressure between 7 and 20 MPa has been reported, with reaction temperature around 270°C [19]. Opposite to the one-step process, under such conditions, no thermal degradation of the product occurs [20]. Some proposals have been reported for industrial-scale biodiesel production flowsheets with a two-step approach [21, 22]. A diagram of a hydrolysis-esterification based production process for biodiesel is shown in Figure 2. For such process, more equipment is required if compared with the one-step process. Nevertheless, the thermal stability of the product is ensured. Moreover, since the first step involves hydrolysis, the presence of water in the raw material has no negative effect on the reaction. Additionally, the hydrolysis step has as objective breaking the triglyceride molecule into free fatty acids. Thus, the presence of such acids in the raw material has no negative effect on the hydrolysis performance. This allows both, the one-step and the two-steps processes, to be good candidates for processing low-quality oils.

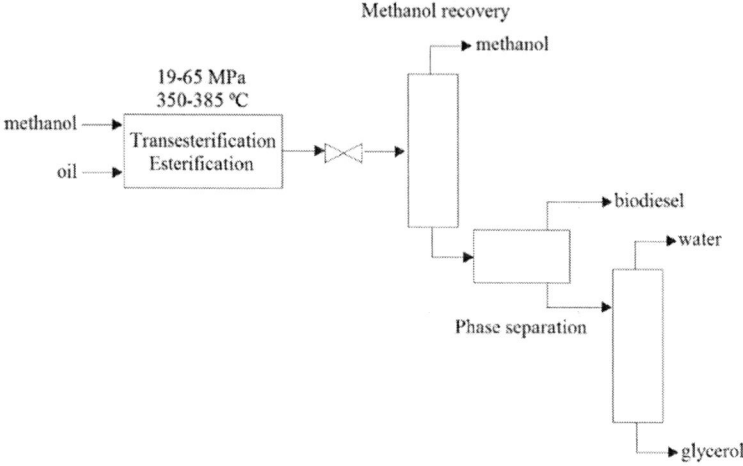

Figure 1. One-step process for the production of biodiesel with supercritical methanol.

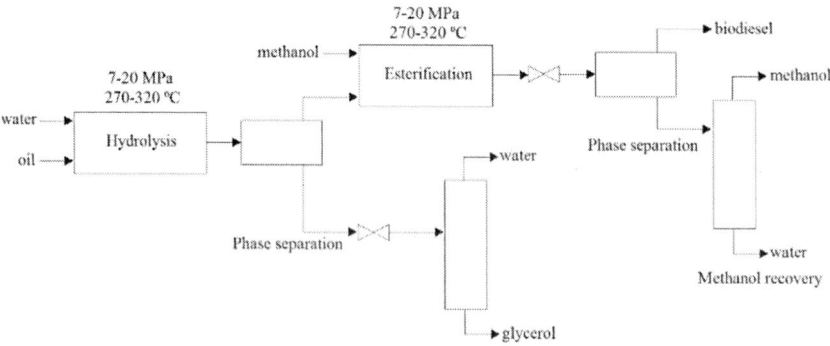

Figure 2. Two-steps process for the production of biodiesel with supercritical methanol.

Most of the biodiesel production processes, including the one-step and the two-steps supercritical processes, have glycerol as a by-product. Nevertheless, some concerns about the falling in the prize of glycerol due to excessive production have been discussed [23]. Saka and Isayama proposed a glycerol-free biodiesel production scheme, using methyl acetate as reactant and obtaining triacetin as a by-product, which can be used as fuel mixed with the produced biodiesel [24], implying that it does not require to be separated from the alkyl esters. Saka and Isayama [24] report that the conditions required for this process are 20 MPa and temperatures in the range of 270-380°C, using rapeseed oil as triglyceride source. Campanelli et al. [25] report similar conditions, 20 MPa and 345°C, for the supercritical treatment of waste cooking oils. Figure 3 shows a simplified flowsheet for an industrial-scale production process of biodiesel with supercritical methyl acetate, based on the proposal reported by Gómez-Castro et al. [15]. In the reactor, free fatty acids can be esterified into alkyl esters, with acetic acid as a by-product. Nevertheless, in the process reported the production of acetic acid is low. Thus no separation process is required to eliminate such product, and there is only need to recover the excess methyl acetate.

The temperature required for the production of biodiesel with supercritical methyl acetate can be higher than 300°C, with the potential of degradation of the biodiesel. As an alternative to avoid high temperatures and simultaneously avoid the production of glycerol as a by-product, Saka et al. [26] proposed a two-step process, where the triglycerides are first transesterified with sub-critical acetic acid, and then the resulting fatty acids are esterified with supercritical methanol. Figure 4 shows a simplified representation of an industrial-scale process for the production of biodiesel using this approach [15]. The process requires more purification steps than other supercritical processes, because of the need to separate the triacetin remaining in the aqueous phase, and the difficulty of separating the binary pair water-acetic acid, implying the need of an entrainer.

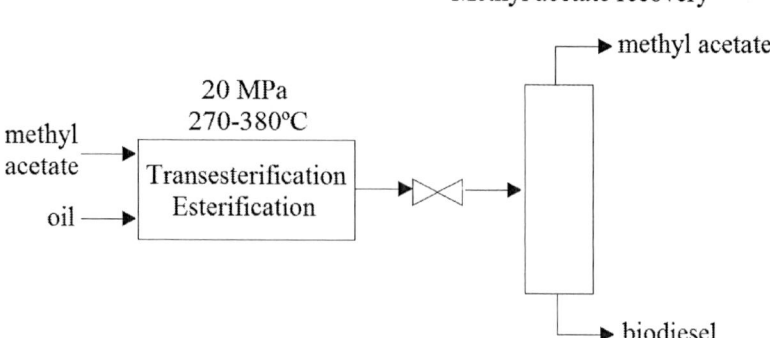

Figure 3. A process for the production of biodiesel with supercritical methyl acetate.

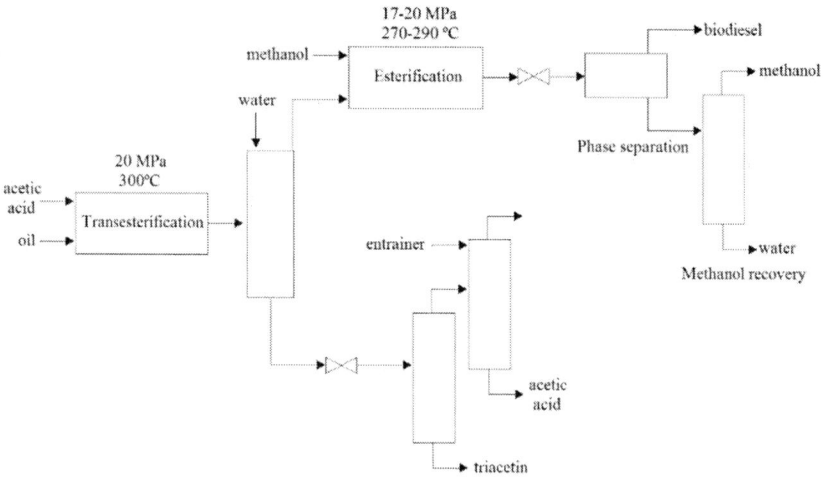

Figure 4. A process for the production of biodiesel with subcritical acetic acid and supercritical methanol.

Another proposal for the non-catalytic production of biodiesel implies the use of dimethyl carbonate, either in a single reaction step [27] or in two steps, with hydrolysis of oils followed by esterification of the resulting fatty acids in supercritical dimethyl carbonate [28]. The one-step system operates under conditions of 15-25 MPa and 320-400°C [27], while the esterification reaction in the two-step approach has been reported for conditions of 6-19 MPa and 270-350°C [28]. To the knowledge of the authors, only preliminary proposals for industrial-scale flowsheets using this approach have been reported [28]. For the detailed analysis of processing alternatives, additional knowledge on the interaction between glyoxal and the other components involved in the processes is required.

As a way to reduce the conditions required in the supercritical processes, but maintaining their main advantages, the use of co-solvents has been proposed and studied by several authors. Han et al. [29] reported the production of biodiesel in a batch reactor using carbon dioxide as co-solvent, with reaction conditions of 14.3 MPa and 280°C. Yin et al. [30] studied the synthesis of biodiesel with supercritical methanol and hexanol or carbon dioxide as co-solvent, finding a high yield of biodiesel at 300°C. Sarve et al. [31] reported the production of biodiesel with supercritical ethanol using carbon dioxide as co-solvent, reporting as best reaction conditions 300°C and 4 MPa of initial CO_2 pressure. Tobar and Núñez [32] presented a study of the transesterification of a microalgae oil in methanol and ethanol with carbon dioxide as co-solvent, reporting the highest yields for 20 MPa and 300°C, using methanol as alcohol. Nevertheless, it is likely that the selected conditions for ethanol were not appropriate. Recently, Akkarawatkhoosith et al. [33] reported the production of biodiesel with supercritical ethanol, using iso-propanol as co-solvent. It is important to recall that iso-propanol can also react to produce alkyl esters. The best reaction conditions reported are 8 MPa and 360°C, obtaining a yield of 97%. This reported pressure is considerably lower than those reported for other supercritical systems, although the temperature is higher. Finally, an interesting analysis of the co-solvent system has been presented by Saka et al. [34], reporting that the highest contribution to the biodiesel yields in such systems is due to the effect of the pressure and the concentration of the reactants, not to the co-solvent itself.

REDUCING ENERGY REQUIREMENTS: PROCESS INTENSIFICATION AND PROCESS INTEGRATION

According to the work of Gómez-Castro et al. [15], conditioning the reactants to the temperature and pressure required by a supercritical reactor could represent around 40-80% of the total energy requirement of an industrial-scale process. Thus, reducing such energy demand is necessary to enhance the economy on the production of biodiesel through such approaches, and to reduce their environmental impact. Making better use of the energy inside of the process is an alternative with the potential to reduce the requirements of external heating, and can be achieved through energy integration studies. Moreover, some of the processes can be enhanced through process intensification, i.e., allowing various unit operations to take place in a single vessel, or making better use of the internal energy flows inside the equipment. Through process intensification, not only energy requirements can be reduced, but also investment costs. Anitescu et al. [35] proposed taking advantage on the reaction conditions to produce electrical power in a cogeneration scheme, through the combustion of a fraction of the stream leaving the reactor in a power generator [36]. Diaz et al. [37] proposed the use of a heat pump in the supercritical production of biodiesel, obtaining thermal energy savings when compared with the process without the heat pump, particularly in a system using propane as co-solvent. Gutiérrez Ortiz and de Santa-Ana [38] proposed using a part of the produced biodiesel to satisfy the energy requirements in a process with propane as co-solvent, also using the hot utility water for cogeneration. Energy integration of the one-step supercritical ethanol process with other biofuel production processes has also been reported. Gómez-Castro et al. [17] presented a proposal of integration with a lignocellulosic bioethanol production process, obtaining energy savings around 4.4% in comparison with the separated processes. Villegas-Herrera et al. [39] proposed the integration with a hydrotreating process for the production of bio-jet fuel, obtaining energy savings of almost 60% in comparison with the non-integrated processes. In terms of process intensification, Gómez-Castro et al. [22] proposed the use of a reactive distillation column to perform the reaction and a preliminary purification in the two-step supercritical methanol process, reporting reductions in utility costs around 15%. Additionally, May-Vázquez et al. [40] studied the hydraulic feasibility of such high-pressure system.

Although there are only few proposals of process intensification and process integration for supercritical approaches in the production of biodiesel, the reports indicate that such schemes have high integration potential, due to the temperature levels required for the reactions. As aforementioned, the supercritical processes need high energy inputs to conditioning the reactants, thus a better use of the availabe energy is mandatory. Integrating such processes in a biorefinery scheme could be an strategy to dramatically reduce the external energy requirements, and consequently reducing the environmental impact.

SAFETY ISSUES IN BIODIESEL PROCESSES

Safety is an essential concept in the chemical industry, and biodiesel production is not an exception. Sometimes, people and companies are not aware of the risk, and unfortunately, they act after a catastrophic accident has already occurred. The risk is evolutionary, and when new processes and technologies arise, also new threats appear. To avoid that biodiesel production, and other biofuel processes, repeat the tragic history of traditional processes, safety must be included in every single stage of the biodiesel facility installation project. Thinking that the biodiesel process is risk-free is a mistake. Some statistics about this emerging industry must prove this statement. A recent study showed that in just one decade (2003 - 2013) 85 events have been registered in biodiesel production plants, including spill or release, fires, and explosions; together with some occupational incidents (slips, falls, burnings and cutting). The primary type of accidents at biodiesel facilities are fire, with 60% of frequency, the explosions representing 30%, while occupational incidents represent only almost 3% of the total of data analyzed. About the consequences, it is important to note that in 19% of the accidents the total loss of the building occurs, in 14% of these events deceases occurred, in 20% there were injured people and significant structural damage. Only 14% of the reported accidents had minor structural damage and there was not injured people, and only 4% of these events did not have consequences at all. The accidents are not exclusive of one region, in all the countries where biodiesel production exists, accidents had occurred. The United State of America (U.S.A.) lead the statistics because it is the most significant world biodiesel producer. Another important point is that most of the accidents occurred during operation [41, 42].

The previous data allow understanding that it is essential to pay attention to safety issues on biodiesel production. To understand the risk of noncatalytic processes it is necessary to remind the definition of risk. According to CCPS the risk is the result of the product between consequences and frequency [43]. Therefore, some aspects of the processes affect the consequence and others the frequency. In the case of noncatalytic processes, there are several features that could increase consequences, e.g., these processes work at high pressure (7–25 MPa), and high pressure propitiates many flammable substances in case of loss of containment. Furthermore, the temperature used in these processes (243 – 380°C) increase the formation rate of the vapor cloud, in case of a flammable substance release scenario. At the same time, the hot surfaces can act as an ignition source. Operation conditions are not the only variables affecting the consequences; the process simplicity itself and the inventory play a significant role in this contribution to the risk. On the other hand, a suitable selection of layers of protection, a good mechanical design, as well as the Safety Integrity Level (SIL) of electronic control devices are determinants in the frequency [44, 45]. Then, the risk of the noncatalytic process cannot be defined in a simple way.

The nature of the substances involved in the noncatalytic biodiesel process is important; the property which can define the type of accident is the flash point. The value of the flash point for a given substance is an indicator of how easily it can vaporize to form an ignitable mixture in air; this implies that the lower the flash point is, the flammability is higher. For a high flash point, the risk of fire prevails, while for a low flash point, the risk of explosion is dominant. Under this definition, all equipment and pipes that transport, produce or storage biodiesel and glycerol, could have a fire, the reason is that they have high flash point value (100°C and 199°C, respectively). On the other hand, methanol and methyl acetate have a low flash point (12°C and -10°C respectively) therefore the pipes, equipment, and tanks with these substances have the potential of fire and explosion.

STRATEGIES TO EVALUATE SAFETY

To reduce the risk in the biodiesel production, then making the plants safer, it is necessary to apply a risk evaluation, which gives information for the optimization of layout, emergency plans, and potential damage to the residential neighbors, especially schools, hospital, and services. The international standard IEC/ISO 31010:2009 presents an excellent general vision of risk analysis tools, including some examples for several engineering disciplines [46]. In general, the strategies to reduce the risk in the processes could be classified into four categories [47]:

1. Inherent: to reduce and eliminate the significant risk
2. Passive: reducing the consequence or probability of incidents through devices that do not require an initial detection; it is put in action by any person or physical device, e.g., rupture disk, relief valve, among others.
3. Active: reducing the consequence or probability of incident through devices that require an initial detection; it is put in action sometimes by any person, but generally are activated automatically, e.g., spray, gas detectors, among others.
4. Procedure: reducing the consequence or probability of incidents due to early detection of an incident, followed by implementation of procedures or devices activated by workers to interrupt the events sequence that results in an incident.

Nowadays, the inherent strategy is the most applied and most effective approach to design a new process, the target of this philosophy is to avoid the hazard or to reduce the probability of accident since the first steps of the project. The other three strategies are implemented once the plant has been installed [48]. However, inherently safety design (ISD) considers the four strategies of risk management.

Different methods have been used to evaluate inherent security; all of them are based on the Kletz principles [49]; some strategies of inherent safety are summarized in Table 1. The methods based on indexing, graphics, and optimization are the most used strategies to evaluate safety. The indexing methodology is usually preferable for comparison of alternative process routes for the same chemical product, this method has excellent flexibility and is appropriate for the first stages of design when less information is available to choose the best process route [50].

Several indexes based on inherent safety have been proposed, the first one was proposed by Edwards and Lawrence, and is known as Prototype Index for Inherent Safety (PIIS) [51]. PIIS is mainly used to analyze the raw materials used and the sequence of the reaction steps. This method focuses on the reaction and is not suitable for the safety analysis of the entire process plant. An extension of PIIS, named Inherent Safety Index (ISI) was presented by Heikkilä et al. [52], this index requires less information compared to other methods while covering many aspects of safety, it considers the nature of the chemical substances and the operation conditions of the process. The limitation of this index is that more detailes information about the process is needed.

Table 1. Strategies used in the inherent safety design

Strategy	Action	Target
Intensification	Minimization	Less amount of hazardous material, and to reduce the equipment size
Attenuation	Moderation	Modify or reduce the operation condition
Substitution	Substitution	Applicable to the material, chemical or process, using substances and process safer
Simplification	Simplification	Eliminate complication from the design
Limitation of effect	Minimization	Change the design or conditions
Flexibility to error	Simplification	Improve the design of equipment to support variations

To evaluate processes in a qualitative way, various indexes have been reported, one of the most complete is the I2SI proposed by Khan and Amyotte [53]. This index is a compilation of previous methodologies presented for the same authors [54, 55], it considers the type of equipment (storage, physical operations, chemical reaction, and transport), the type of accident (fire, explosion, toxic releases), and finally other hazardous units as flare and oven. A complete review of the indexes available to perform qualitative and quantitative risk analysis was presented by Roy Nitin et al. [56]. Other methodologies that are not based on inherent safety, but continues in use due to its effectivity, is the Dow Fire & Explosion Index (F&EI), and the Mond Index, which are well known and widely used in process industries [57, 58]. Both methodologies were created by Dow Chemical Company, summarizing a long time of knowledge and experience. Both are valuable tools since the indexes identify which equipment must be considered hazardous and where the installation of additional protection devices could be necessary. However, both indexes are more useful in an intermedium stage of the design, because more detailed information about the process is necessary, e.g., size of equipment, plant layout, installed control devices, passive safeguards, among others.

Probably the most popular method to identify hazards in the industries is the Hazard and Operability Analysis (HAZOP). This method starts as an operability studio of the new process, but at the same time, the method helps to identify and analyze risk in the process. Nowadays, HAZOP is a methodology based on an expert team with different specialization areas. The team discuss together all the possible failures and their consequences, and at the same time define the safeguards and modification to improve the process safety. There is a large amount of literature about HAZOP; which is a quantitative method [59]. The use of this methodology is recommended from the medium level of design because the information required is more specific to perform a proper analysis. It is important to remark that all the mentioned methods have been developed and probed in chemical and petrochemical processes, but some of them has been applied in the biodiesel production. The next section will comment on some examples of safety analysis in biodiesel production as well as some opportunities in this area.

SAFETY EVALUATION OF BIODIESEL PRODUCTION

Concerning about safety in the biodiesel process plants has increased in recent years and specific methodologies are proposed continuously. However, several methodologies used in the chemical processes could be adapted for this kind of facilities. The proper methodology depends on the project stage: for a project on the first stages, a process hazard identification is recommended, as well as a semi quantitative or qualitative analysis based on inherent safety. There are two adaptations for this biodiesel project stage. The first one is a modification of Inherent Safety Index; the original ISI method does not count the complexity of the process or the quantity of chemicals used in the process. So, a research team proposed the Enhanced Inherent Safety Index (EISI) method to enhance its functionality, the application of this index was presented with the incorporation of exergy analysis for biodiesel production [60]. Other application of this index can be reviewed in the work of Jayswal et al. [61], who performed a sustainability analysis assuming the safety of biodiesel process as part of the social impact. Another approach which claims to be useful on the initial stage of biodiesel production is a modification of EISI. This index is named the comprehensive inherent safety index (CISI). It adopts an object-oriented approach for inherent safety analysis. Just like the object-oriented approach in software programming, a significant advantage of the CISI is that it provides a clear modular structure, where each equipment forms a separate entity with its own safety score [62]. On the other hand, if the project is at an advanced stage, even at the last stage, the best methodologies to evaluate and identify hazard are the Dow F&E Index, the Mod index or, the best alternative to the date, the HAZOP analysis. It is important to remark that the above-presented analysis was performed for catalytic processes. There is no report for the production of biodiesel using alcohols at supercritical conditions. Therefore, safety in noncatalytic and intensified processes is an emerging field of opportunities. Some characteristics that make this kind of processes interesting, from the safety point of view are: in some processes the number of equipment are reduced, leading less pipework (reducing the sources of leaks); some supercritical biodiesel processes are simpler than the conventional process, and simplification is an inherent strategy; in some cases, the amount of hazardous material is reduced, so smaller vessel is required and the same time, these vessels are easier to design and control. In some cases, safety can be benefited from process intensification (PI), when the number of equipment is reduced [63]. Nevertheless, in other systems the intensification may lead to potential safety problems, which may include: some intensified systems require high-energy inputs, and some companies could not have experience handling high-energy sources, then these conditions could trigger new hazards; the processes may need more complex control systems, having a negative impact on safety; the high-energy sources may introduce new hazards that have to be considered when applied to hazardous substances, e.g., whether or not it is safe to use microwaves on thermally unstable substances or mixtures; in some cases, process pipework may be complex with a higher potential for equipment failure or operator error. Another important aspect is the plant layout design; it has been proven that layout plays an important role in process safety [64]. Even the selection of a proper layout can be considered as a strategy of inherent safety design. This part of the project is not a simple task and plays a relevant role in the consequences of an accident. To facilitate these task, some mathematical approaches have been presented to optimize the layout, considering safety issue, particularly explosion scenarios [44, 65]. These models can help to improve the inherent safety design.

CONCLUSION

The production of biodiesel can be performed through supercritical approaches, having high potential to treat low-cost raw materials. In terms of total annual costs, the one-step process and the methyl acetate process seem to be the best alternative, but they may be limited because of the associated environmental impact. The use of tools as process intensification and process integration may help to reduce such impact, integrating the supercritical processes with other biofuel production plants. The use of co-solvents can also be an alternative to reduce the harsh temperature conditions of the reaction, but further research is necessary.

It can be observed that defining the risk of the noncatalytic biodiesel processes is not a simple task. Most of the proposed methodologies are focused on just one component of the risk, i.e., the consequences. The available methodologies used to identify hazards in biodiesel production still have limitations; then it is necessary to continue improving the strategy to identify hazards, since the use of a reliable tool is essential to choose the safer design. The safety on biofuel production processes must be considered seriously, and this philosophy will make the difference between the dangerous plants built in the past, and the new generation of safe, cleaner processes that satisfy the challenges of the future.

REFERENCES

[1] Dickey, P. A. (1959). The first oil well. *Journal of Petroleum Technology*, 11(1): 14-25.

[2] Chapman, I. (2014). The end of Peak Oil? Why this topic is still relevant despite recent denials. *Energy Policy*, 64 (January): 93-101.

[3] British Petroleum, *Statistical Review of World Energy*, in https://www.bp.com/en/global/corporate/energy-economics/statistical-review-of-world-energy.html. Last visited 11 February 2019.

[4] Environmental Protection Agency, *Climate Change Indicators: Global Greenhouse Gas Emissions*, in https://www.epa.gov/climate-indicators/climate-change-indicators-global-greenhouse-gas-emissions. Last visited 12 February 2019.

[5] Vicente, G., Martínez, M. and Aracil, J. (2004). Integrated biodiesel production: a comparison of different homogeneous catalysts systems. *Bioresource Technology*, 92(3): 297-305.

[6] Gasca-González, R., Gómez-Castro, F. I., Romero-Izquierdo, A. G., Zenón-Olvera, E. and Gutiérrez-Antonio, C. (2018). Design of a low-cost process for the production of biodiesel using waste oil as raw material, *Computer Aided Chemical Engineering*, 43: 1529-1534.

[7] Lee, S., Posarac, D. and Ellis, N. (2011). Process simulation and economic analysis of biodiesel production process using fresh and waste vegetable oil and supercritical methanol. *Chemical Engineering Research and Design*, 89(12): 2626-2642.

[8] Glisic, S. and Skala, D. (2009). The problems in design and detailed analyses of energy consumption for biodiesel synthesis at supercritical conditions. *The Journal of Supercritical Fluids*, 49(2): 293-301.

[9] Center for Chemical Process Safety (2008). *Inherently safer chemical processes: a life cycle approach*. John Wiley and Sons, New Jersey.

[10] Kusdiana, D. and Saka, S. (2001). Kinetics of transesterification in rapeseed oil to biodiesel fuel as treated in supercritical methanol. *Fuel*, 80(5): 693-698.

[11] Saka, S. and Kusdiana, D. (2001). Biodiesel fuel from rapeseed oil as prepared in supercritical methanol. *Fuel*, 80(2): 225-231.

[12] Kusdiana, D. and Saka, S. (2004). Effect of water on biodiesel fuel production by supercritical methanol treatment, *Bioresource Technology*, 91(3): 289-295.

[13] Warabi, Y., Kusdiana, D. and Saka, S. (2004). Reactivity of triglycerides and fatty acids of grapeseed oil in supercritical alcohols, *Bioresource Technology*, 91(3): 283-287.

[14] Lee, S., Posarac, D. and Ellis, N. (2011). Process simulation and economic analysis of biodiesel production processes using fresh and waste vegetable oil and supercritical methanol, *Chemical Engineering Research and Design*, 89(12): 2626-2642.

[15] Gómez-Castro, F. I., Segovia-Hernández, J. G., Hernández, S., Rico-Ramírez, V., Gutiérrez-Antonio, C., Briones-Ramírez, A., Cano-

Rodríguez, I. and Gamiño-Arroyo, Z. (2015). Analysis of alternative non-catalytic processes for the production of biodiesel fuel, *Clean Technologies and Environmental Policy*, 17(7): 2041-2054.

[16] Gui, M. M., Lee, K. T. and Bhatia, S. (2009). Supercritical ethanol technology for the production of biodiesel: Process optimization studies, *The Journal of Supercritical Fluids*, 49(2): 286-292.

[17] Gómez-Castro, F. I., Aldana-González, M. G., Conde-Mejía, C., Gutiérrez-Antonio, C., Romero-Izquierdo, A. G. and Morales-Rodríguez, R. (2017). Process integration for the supercritical production of biodiesel and the production of lignocellulosic bioethanol, *Computer Aided Chemical Engineering*, 40: 931-936.

[18] Saka, S. (2005). Biodiesel fuel production by supercritical methanol technology, *Journal of the Japan Institute of Energy*, 84(4): 413-149.

[19] Minami, E. and Saka, S. (2006). Kinetics of hydrolisis and methyl esterification for biodiesel production in two-step supercritical methanol process, *Fuel*, 85(17): 2479-2483.

[20] Imahara, H., Minami, E., Hari, S. and Saka, S. (2008). Thermal stability of biodiesel in supercritical methanol, *Fuel*, 87(1): 1-6.

[21] D'Ippolito, S. A., Yori, J. C., Iturria, M. E., Pieck, C. L. and Vera, C. R. (2007). Analysis of a two-step, noncatalytic, supercritical biodiesel production process with heat recovery, *Energy & Fuels*, 21(1): 339-346.

[22] Gómez-Castro, F. I., Rico-Ramírez, V., Segovia-Hernández, J. G., Hernández-Castro, S. and El-Halwagi, M. M. (2013). Simulation study on biodiesel production by reactive distillation with methanol at high pressure and temperature: Impact on costs and pollutant emissions, *Computers and Chemical Engineering*, 52(May): 204-215.

[23] Johnson, D. T. and Taconi, K. A. (2007). The glycerin glut: options for the value-added conversion of crude glycerol resulting from biodiesel production, *Environmental Progress*, 26(4): 338-348.

[24] Saka, S. and Isayama, Y. (2009). A new process for catalyst-free production of biodiesel using supercritical methyl acetate, *Fuel*, 88(7): 1307-1313.

[25] Campanelli, P., Banchero, M. and Manna, L. (2010). Synthesis of biodiesel from edible, non-edible and waste cooking oils via supercritical methyl acetate transesterification, *Fuel*, 89(12): 3675-3682.

[26] Saka, S., Isayama, Y., Ilham, Z. and Jiayu, X. (2010). New process for catalyst-free biodiesel production using subcritical acetic acid and supercritical methanol, *Fuel*, 89(7): 1442-1446.

[27] Tan, K. T., Lee, K. T. and Mohamed, A. R. (2010). Optimization of supercritical dimethyl carbonate (SCDMC) technology for the production of biodiesel and value-added glycerol carbonate, *Fuel*, 89(12): 3833-3839.

[28] Ilham, Z. and Saka, S. (2010). Two-step supercritical dimethyl carbonate method for biodiesel production from *Jatropha curcas* oil, *Bioresource Technology*, 101(8): 2735-2740.

[29] Han, H., Cao, W. and Zhang, J. (2005). Preparation of biodiesel from soybean oil using supercritical methanol and CO_2 as co-solvent, *Process Biochemistry*, 40(9): 3148-3151.

[30] Yin, J. Z., Xiao, M. and Song, J. B. (2008). Biodiesel from soybean oil in supercritical methanol with co-solvent, *Energy Conversion and Management*, 40(5): 908-912.

[31] Sarve, A. N., Varma, M. N. and Sonawane, S. S. (2015). Response surface optimization and artificial neural network modeling of biodiesel production from crude mahua (*Madhuca indica*) oil under supercritical ethanol conditions using CO_2 as co-solvent, *Royal Society of Chemistry Advances*, 5: 69702-69713.

[32] Tobar, M. and Núñez, G. A. (2018). Supercritical transesterification of microalgae triglycerides for biodiesel production: Effect of alcohol type and co-solvent, *The Journal of Supercritical Fluids*, 137(July): 50-56.

[33] Akkarawatkhoosith, N., Kaewchada, A. and Attasak, J. (2019). Simultaneous development of biodiesel synthesis and fuel quality via continuous supercritical process with reactive co-solvent, *Fuel*, 237(February) : 117-125.

[34] Imahara, H., Xin, J. and Saka, S. (2009). Effect of CO_2/N_2 addition to supercritical methanol on reactivities and fuel qualities in biodiesel production, *Fuel*, 88(7): 1329-1332.

[35] Anitescu, G., Desphande, A. and Tavlarides, L. L. (2008). Integrated technology for supercritical biodiesel production and power cogeneration, *Energy and Fuels*, 22(2) : 1391-1399.

[36] Desphande, A., Anitescu, G., Rice, P. A. and Tavlarides, L. L. (2010). Supercritical biodiesel production and power cogeneration: Technical and economic feasibilities, *Bioresource Technology*, 101(6) : 1834-1843.

[37] Diaz, M. S., Espinosa, S. and Brignole, E. A. (2009). Model-based cost minimization in noncatalytic biodiesel production plants, *Energy and Fuels*, 23(11) : 5587-5595.

[38] Gutiérrez Ortiz, F. J. and de Santa-Ana, P. (2017). Techno-economic assessment of an energy self-sufficient process to produce biodiesel under supercritical conditions, *The Journal of Supercritical Fluids*, 128(October) : 349-358.

[39] Villegas-Herrera, L. A., Gómez-Castro, F. I., Romero-Izquierdo, A. G., Gutiérrez-Antonio, C. and Hernández, S. (2018). Feasibility of energy integration for high-pressure biofuels production processes, *Computer Aided Chemical Engineering*, 43: 1523-1528.

[40] May-Vázquez, M. M., Rodríguez-Ángeles, M. A., Gómez-Castro, F. I. and Uribe-Ramírez, A. R. (2017). Hydrodynamic feasibility of the production of biodiesel in a high-pressure reactive distillation column, *Chemical Engineering and Processing: Process Intensification*, 112(February): 31-37.

[41] Calvo Olivares, R. D., Rivera, S. S. and Núñez Mc Leod, J. E. (2014). Database for accidents and incidents in the biodiesel industry. *Journal of Loss Prevention in the Process Industries*, 29: 245-261.

[42] Casson Moreno, V. and Cozzani, V. (2015). Major accident hazard in bioenergy production. *Journal of Loss Prevention in the Process Industries*, 35: 135-144.

[43] Freeman, R. A. (1990). CCPS guidelines for chemical process quantitative risk analysis. *Plant/Operations Progress*, 9(4): 231-235.

[44] de Lira-Flores, J. A., López-Molina, A., Gutiérrez-Antonio, C. and Vázquez-Román, R. (2019). Optimal plant layout considering the safety instrumented system design for hazardous equipment. *Process Safety and Environmental Protection*, *124*: 97-120.

[45] Smith, D. J. and Simpson, K. G. (2016). *The Safety Critical Systems Handbook: A Straightforward Guide to Functional Safety: IEC 61508 (2010 Edition), IEC 61511 (2015 Edition) and Related Guidance.* Butterworth-Heinemann.

[46] IEC, I. (2010). ISO 31010: 2009: *Risk Management - Risk Assessment Techniques, vol. 2010E.* Brussels. CENELEC.

[47] CCPS. (1996). *Guidelines for Use of Vapor Cloud Dispersion Models.* Second edition. CCPS-Wiley.

[48] Leong, C. T. and Shariff, A. M. (2009). Process route index (PRI) to assess level of explosiveness for inherent safety quantification. *Journal of Loss Prevention in the Process Industries*, *22*(2): 216-221.

[49] Kletz, T. A. and Amyotte, P. (2010). *Process plants: A handbook for inherently safer design.* CRC Press.

[50] Srinivasan, R. and Nhan, N. T. (2008). A statistical approach for evaluating inherent benign-ness of chemical process routes in early design stages. *Process Safety and Environmental Protection*, *86*(3): 163-174.

[51] Edwards, D. W. and Lawrence, D. (1993). Assessing the inherent safety of chemical process routes: is there a relation between plant costs and inherent safety? *Process Safety and Environmental Protection*, 71: 252-258.

[52] Heikkilä, A. M., Hurme, M. and Järveläinen, M. (1996). Safety considerations in process synthesis. *Computers & Chemical Engineering*, 20: S115-S120.

[53] Khan, F. I. and Amyotte, P. R. (2005). I2SI: A comprehensive quantitative tool for inherent safety and cost evaluation. *Journal of Loss Prevention in the Process Industries*, 18(4): 310-326.

[54] Khan, F. R. and Amyotte, P. (2004). Integrated inherent safety index (I2SI), A tool for inherent safety evaluation. *Process Safety Progress*, 23(2): 136-148.

[55] Khan, F. I., Husain, T. and Abbasi, S. A. (2001). Safety weighted hazard index (SWeHI): A new, user-friendly tool for swift yet comprehensive hazard identification and safety evaluation in chemical process industrie. *Process Safety and Environmental Protection*, 79(2): 65-80.

[56] Roy, N., Eljack, F., Jiménez-Gutiérrez, A., Zhang, B., Thiruvenkataswamy, P., El-Halwagi, M., Mannan, M.S. (2016). A review of safety indices for process design. *Current Opinion in Chemical Engineering*, 14(November): 42-48.

[57] Dow. (1994). *Dow Fire and Explosion Index hazard Classification Guide*. 7th ed. ed., New York: American Institute of Chemical Engineers.

[58] Dow. (1994). Dow's Chemical Exposure Index. New York, AIChE.

[59] Dunjó, J., Fthenakis, V., Vílchez, J. A. and Arnaldos, J. (2010). Hazard and operability (HAZOP) analysis. A literature review. *Journal of Hazardous Materials*, 173(1-3): 19-32.

[60] Li, X., Zanwar, A., Jayswal, A., Lou, H. H. and Huang, Y. (2011). Incorporating exergy analysis and inherent safety analysis for sustainability assessment of biofuels. *Industrial & Engineering Chemistry Research*, 50(5): 2981-2993.

[61] Jayswal, A., Li, X., Zanwar, A., Lou, H. H. and Huang, Y. (2011). A sustainability root cause analysis methodology and its application. *Computers & Chemical Engineering*, 35(12): 2786-2798.

[62] Gangadharan, P., Singh, R., Cheng, F. and Lou, H. (2013). Novel Methodology for Inherent Safety Assessment in the Process Design Stage. *Industrial & Engineering Chemistry Research*, 52(17): 5921–5933.

[63] Contreras-Vargas, C. A., Gómez-Castro, F. I., Sánchez-Ramírez, E., Segovia-Hernández, J. G., Morales-Rodríguez, R. and Gamiño-Arroyo, Z. (2019). Alternatives for the purification of the blend butanol/ethanol from an acetone/butanol/ethanol fermentation effluent. *Chemical Engineering & Technology*, 42(5): 1088-1100.

[64] Mannan, S. (2005). *Lees' loss prevention in the process industries: Hazard identification, assessment and control.* USA, Elsevier Butterworth-Heinemann.

[65] López-Molina, A., Vázquez-Román, R., Mannan, M. S. and Félix-Flores, M. G. (2013). An approach for domino effect reduction based on optimal layouts. *Journal of Loss Prevention in the Process Industries*, 26(5): 887-894.

BIOGRAPHICAL SKETCHES

Fernando Israel Gómez Castro

Affiliation: Departamento de Ingeniería Química, División de Ciencias Naturales y Exactas, Campus Guanajuato, Universidad de Guanajuato

Education: ScD in Chemical Engineering

Research and Professional Experience: Researcher in the area of process design, analysis and optimization, recognized as National Researcher by the National System of Researchers (SNI, México). 37 scientific papers published in indexed journals, 7 book chapters and three books. Participation in several national and international scientific conferences. Reviewer for different indexed journals. Current Head of the Bachelor Program on Chemical Engineering of the Universidad de Guanajuato.

Professional Appointments: Member of the Directive Board of the Mexican Academy of Research and Teaching in Chemical Engineering (AMIDIQ). Member of the America Chemical Society. Registered as accredited evaluator for the National Council of Science and Technology (Mexico). Member of the Thematic Network on Bioenergy (Mexico).

Honors: Elected as Member of the Directive Board of the Mexican Academy of Research and Teaching in Chemical Engineering (AMIDIQ) in the periods 2017-2019 and 2019-2021. 2017 Certificate of Excellence in Reviewing (Clean Technologies and Environmental Policy). 2018 Certificate of Outstanding Contribution in Reviewing (Applied Soft Computing).

Publications from the Last 3 Years:

Alfaro-Ayala, J. A., López-Núñez, O. A., Gómez-Castro, F. I., Ramírez-Minguela, J. J., Uribe- Ramírez, A. R., Belman-Flores, J. M. and Cano-Andrade, S. (2018). Optimization of a solar collector with evacuated tubes using the simulated annealing and computational fluid dynamics. *Energy Conversion and Management*, 166(June): 343-355.

Arce-Alejandro, R., Villegas-Alcaraz, J.F., Gómez-Castro, F.I., Juárez-Trujillo, L., Sánchez- Ramírez, E., Carrera-Rodríguez, M. and Morales-Rodríguez, R. (2018). Performance of a gasoline engine powered by a mixture of ethanol and n-butanol. *Clean Technologies and Environmental Policy*, 20: 1929-1937.

Contreras-Vargas, C. A., Gómez-Castro, F. I., Sánchez-Ramírez, E., Segovia-Hernández, J. G., Morales-Rodríguez, R. and Gamiño-Arroyo, Z. (2019). Alternatives for the purification of the blend butanol/ethanol from an ABE fermentation effluent: impact on the economic, environmental and safety indexes. *Chemical Engineering & Technology*, 42(5): 1088-1100.

Gómez-Castro, F. I. and Segovia-Hernández, J. G., Eds. (2019) *Process Intensification: Design Methodologies.* De Gruyter, Germany.

Gómez-Castro, F. I., Gutiérrez-Antonio, C., Briones-Ramírez, A. and Segovia-Hernández, J. G. (2017). Genetic algorithms: A tool for optimizing intensified distillation sequences, in *Genetic Algorithms Advances in Research and Applications*, Edited by J. Carlson, Nova Science Publishers: 1-17.

Gutiérrez-Antonio, C., Gómez-Castro, F. I., de Lira-Flores, J. A. and Hernandez, S. (2017). A review on the production processes of renewable jet fuel. *Renewable & Sustainable Energy Reviews*, 79 (November): 709-729.

Gutiérrez-Antonio, C., Gómez-De la Cruz, A., Romero-Izquierdo, A.G., Gómez-Castro, F. I. and Hernández, S. (2018). Modeling, simulation and intensification of hydroprocessing of micro-algae oil to produce renewable aviation fuel. *Clean Technologies and Environmental Policy*, 20: 1589-1598.

Gutiérrez-Antonio, C., Soria-Ornelas, M. L., Gómez-Castro, F. I. and Hernández, S. (2018). Intensification of the hydrotreating process to produce renewable aviation fuel through reactive distillation. *Chemical Engineering and Processing: Process Intensification*, 124(February): 122-130.

May-Vázquez, M. M., Rodríguez-Ángeles, M. A., Gómez-Castro, F. I. and Uribe-Ramírez, A. R. (2017). Hydrodynamic feasibility of the production of biodiesel fuel in a high-pressure reactive distillation column. *Chemical Engineering & Processing*, 112(February): 31-37.

Méndez-Vázquez, M. A., Gómez-Castro, F. I., Ponce-Ortega, J. M., Serafín-Muñoz, A. H., Santibañez-Aguilar, J. E. and El-Halwagi, M. M. (2017). Mathematical optimization of a supply chain for the production of fuel pellets from residual biomass, *Clean Technologies and Environmental Policy*, 19: 721-734.

Páramo-Vargas, J., Maldonado-Rubio, M. I., Gómez-Castro, F. I. and Peralta-Hernández, J. M. (2017). Modeling the Fenton depuration of the effluent from a slaughterhouse based on design of experiments. *MOJ Ecology & Environmental Science*, 2(2), 00018.

Quintero-Almanza, D., Gamiño-Arroyo, Z., Sánchez-Cadena, L. E., Gómez-Castro, F. I., Uribe- Ramírez, A. R., Aguilera-Alvarado, A. F. and Ocampo Carmona, L. M., (2019). Recovery of cobalt from spent lithium-ion mobile phone batteries using liquid-liquid extraction. *Batteries*, 5(2): 44-57.

Romero-Izquierdo, A. G., Gutiérrez-Antonio, C., Gómez-Castro, F. I. and Hernández, S. (2018). Hydrotreating of triglyceride feedstock to produce renewable aviation fuel. *Recent Innovations in Chemical Engineering*, 11(2): 77-89.

Segovia-Hernández, J. G. and Gómez-Castro, F. I. (2017) *Stochastic Process Optimization Using Aspen® Plus*. CRC Press, Boca Raton, FL.

Segovia-Hernández, J. G., Gómez-Castro, F. I. and Sánchez-Ramírez, E., (2018). Dynamic performance of a complex distillation configuration for the separation of a five-components hydrocarbon mixture. *Chemical Engineering & Technology*, 41(10): 2053-2065.

Segovia-Hernández, J. G., Gómez-Castro, F. I., Vázquez-Castillo, J. A., Contreras-Zarazúa, G. and Gutiérrez-Antonio, C. (2017). Process synthesis and intensification by integration between process design and control, in *Process Synthesis and Process Intensification: Methodological Approaches*, Edited by B.G. Rong, De Gruyter: 370-404.

Velázquez-Guevara, M. A., Uribe-Ramírez, A. R., Gómez-Castro, F. I., Ponce-Ortega, J. M., Hernández, S., Segovia-Hernández, J. G., Alfaro-Ayala, J. A. & Ramírez-Minguela, J. J. (2018). Synthesis of mass exchange networks: a novel mathematical programming approach. *Computers and Chemical Engineering*, 115(July): 226-232.

Antioco López Molina

Affiliation: Ingeniería Petroquímica, División Académica Multidisciplinaria de Jalpa de Méndez, Universidad Juárez Autónoma de Tabasco

Education: ScD in Chemical Engineering

Research and Professional Experience: Researcher in the area of process safety, process design, and optimization, recognized as National Researcher by the National System of Researchers (SNI, México). 10 scientific papers published in indexed journals. Participation in several national and international scientific conferences. Reviewer for different indexed journals. Current Head of the Bachelor Program on Petrochemical Engineering of the Universidad Juárez Autónoma de Tabasco.

Professional Appointments: Member of the Mexican Academy of Research and Teaching in Chemical Engineering (AMIDIQ). Registered as accredited evaluator for the National Council of Science and Technology (Mexico).

Publications from the Last 3 Years:

Conde-Báez, L., López-Molina, A., Gómez-Aldapa, C., Pineda-Muñoz, C. and Conde-Mejía, C. (2019). Economic projection of 2-phenylethanol production from whey. *Food and Bioproducts Processing*, 115(May), 10-16.

Conde-Mejía, C., Aguilar-Arteaga, K., López-Molina, A. and Guerrero-Zárate, D. (2019). Producción de biomasa en la ficorremediación de efluentes de la industria láctea, *Dyna Ingeniería e Industria*, 94: 360. [Biomass production in the phycoremediation of effluents from the dairy industry, *Dyna Engineering and Industry*, 94: 360]

de Lira-Flores, J. A., López-Molina, A., Gutiérrez-Antonio, C. and Vázquez-Román, R. (2019). Optimal plant layout considering the safety instrumented system design for hazardous equipment. *Process Safety and Environmental Protection*, 124(April): 97-120.

López-Molina, A., Conde-Mejía, C., Hernández-Martínez, P., Aguilar-Arteaga, K. and Rivera-Aguilar K. Z. (2018). Microalgae harvesting from wastewater by electroflocculation: source for biofuel production, *Recent Innovations in Chemical Engineering*, 11(2), 90-98.

In: Biofuels
Editor: George R. Carey
ISBN: 978-1-53617-721-3
© 2020 Nova Science Publishers, Inc.

Chapter 2

THERMODYNAMIC PROPERTIES OF BIOFUELS: COMPARISON AND REVIEW OF EXCESS ENTHALPY OF MIXTURES OF BUTANOL, OR DIBUTYLETHER, WITH REPRESENTATIVE HYDROCARBONS

F. Aguilar[1], PhD, N. Muñoz-Rujas[1], PhD, E. Montero[1],*, PhD and F. E. M. Alaoui[2], PhD

[1]Department of Electromechanical Engineering,
Universidad de Burgos, Burgos, Spain
[2]Ecole Nationale des Sciences Appliquées,
Université Chouaib Doukkali, El Jadida, Morocco

ABSTRACT

The use of alcohols and ethers as oxygenated compounds in gasoline blends has been proposed in order to reduce the emmisions of new reformulated gasoline. Alcohol, or ether + hydrocarbon mixtures are of interest as model mixtures for gasoline in which the alcohol and the ether act as non-polluting, high octane number blending agents.

---

* Corresponding Author's E-mail: emontero@ubu.es.

Higher alcohols (those containing more than two carbon atoms) coming from renewable sources, and several other oxygenated compounds, can be used as blend components in gasoline for the reduction of petroleum consumption and greenhouse gas emissions. Amongst these oxygenates, butanol is considered as an alternative to conventional gasoline and diesel fuels. Butanol has many advantages over other potential alternative fuel candidates such as ethanol. Butanol can be demonstrated to work in the internal combustion engines designed for use with gasoline without modification at a composition rate of 85% in volume (unlike 85% ethanol, E85). Butanol presents a similar contribution to the antiknock effect to those of methanol and ethanol, while its energy content per unit volume is higher than ethanol, and almost as high as gasoline. Butanol is less susceptible to separation in the presence of water than ethanol/gasoline blends, therefore allowing the use of the existing distribution infrastructure without requiring modifications to blending facilities, storage tanks or retail station pumps.

1-Butoxybutane, also known as dibutyl ether (DBE) is considered to be a valuable additive to second generation bio-fuels. It acts as a non-polluting, high octane number compound used as a blending agent in reformulated gasoline. And butanol is a basic component in the synthesis of the ether, and therefore is always present as an impurity. Thus, the mixtures of DBE, or butanol, with the hydrocarbons representative of the gasolines is a topic of interest in the modelling and prediction of properties of future biofuels. Moreover, for the aim of optimizing common industrial processes (storage, transport, separation and mixing processes), reliable experimental data are needed. Accurate empirical equations, models and simulation programs need to be fed with such experimental data to be useful.

This work presents a review of the experimental excess enthalpy (or enthalpy of mixing) of mixtures of DBE, or butanol, with representative hydrocarbons. The most relevant functional groups of gasoline hydrocarbon types are considered: heptane (alkane), iso-octane (branched alkane), 1-hexene (alkene), cyclohexane (cyclic), methylcyclohexane (branched cyclic), benzene (aromatic), and toluene (branched aromatic). Excess enthalpy of mixtures is a valuable property when evaluating the mixing and storage behaviour of fuels. The excess enthalpies of a set of 14 binary systems of oxygenated + hydrocarbon experimentally determined by the authors are presented and compared, at 298.15 and 313.15 K. In addition, a literature review on the excess enthalpy of mixtures of 1-butanol, or dibutyl ether, with representative hydrocarbons is included. This review could be of interest for the biofuel industry, within the production, transport and end-user (automotive) sectors.

Keywords: biofuels, butanol, dibutyl ether, enthalpy

INTRODUCTION

Biofuels, as environmental friendly fluids, have been paid much attention over the last decades. They contribute to diminishing greenhouse gas emissions due to their neutral carbon dioxide balance. Moreover, some oxygenated compounds are used as biofuel additives as they lead to a reduction in pollutant emissions and to an increase in the energy efficiency of vehicle engines [1, 2].

Some alcohols and ethers are added to present-day gasolines as oxygenated-compound additives with the aim of reducing the emission of gases that produce environmental impact. The benefits of using these oxygenates are not only environmental. First, they can be obtained from renewable agricultural and raw materials, reducing the dependence of fossil sources [3]. Second, they enhance the octane number, boosting the anti-knock effect in gasoline. Then, the compression ratio of the engines can be increased without the risk of knocking, leading to a higher delivery of power. From the combustion point of view, the production of carbon monoxide and volatile hydrocarbons from the combustion of alcohols is smaller than the one of gasoline. Amongst the thermodynamic properties, the heat of vaporization of alcohols is high and leads to a reduction in the peak temperature of combustion, which means lower emissions of nitrogen oxides.

Amongst these oxygenates, butanol has been proposed as an alternative to conventional gasoline and diesel fuels [4, 5]. Butanol has many advantages over other potential alternative fuel candidates such as ethanol. Butanol can be demonstrated to work in the internal combustion engines designed for use with gasoline without modification at a composition rate of 85% in volume (unlike 85% ethanol, E85). Butanol presents a similar contribution to the antiknock effect to those of methanol and ethanol, while its energy content per unit volume is higher than ethanol, and almost as high as gasoline. Butanol is less susceptible to separation in the presence of water than ethanol/gasoline blends, therefore allowing the use of the existing distribution infrastructure without requiring modifications to blending facilities, storage tanks or retail station pumps.

1-Butoxybutane, also known as dibutyl ether (DBE) is considered to be a valuable additive to second generation bio-fuels. It acts as non-polluting, high octane number compound used as blending agent in reformulated gasoline. And butanol is a basic component in the synthesis of the ether, and therefore is always present as an impurity. Thus, the mixtures of DBE, or butanol, with the hydrocarbons representative of the gasolines is a topic of interest in the modelling and prediction of properties of future biofuels. Moreover, for the aim of optimizing common industrial processes (storage, transport, separation and mixing processes), reliable experimental data are needed. Accurate empirical equations, models and simulation programs need to be fed with such experimental data to be useful.

This work presents a review of the experimental excess enthalpy (or enthalpy of mixing) of mixtures of DBE, or butanol, with representative hydrocarbons. The most relevant functional groups of gasoline hydrocarbon types are considered: heptane (alkane), iso-octane or 2,2,4 trimthylpentane (branched alkane), 1-hexene (alkene), cyclohexane (cyclic), methylcyclohexane (branched cyclic), benzene (aromatic), and toluene (branched aromatic). Excess enthalpy of mixtures is a valuable property when evaluating the mixing and storage behaviour of fuels. The excess enthalpy of a set of 14 binary systems of oxygenated + hydrocarbon experimentally determined by the authors [6-11] are presented and compared, at 298.15 and 313.15 K. This review could be of interest for the biofuel industry, within the production, transport and end-user (automotive) sectors.

Experimental Section

Chemicals

All the chemicals used in the experiments were purchased from Fluka Chemie AG and were of the highest purity available, chromatography quality reagents (of the series puriss p.a.) with a stated purity >99.5 mol%. The purity of all reagents was checked by gas chromatography, and their values are presented in Table 1.

Table 1. Purity and related data of chemicals

Compound	Formula	Molar mass/g·mol^{-1}	Stated purity/mol%
DBE	$C_8H_{18}O$	130.2	>99.6%
1-Butanol	$C_4H_{10}O$	74.12	>99.9%
Heptane	C_7H_{14}	100.2	>99.8%
Iso-octane	C_8H_{18}	114.2	>99.9%
1-Hexene	C_6H_{12}	84.16	>99.3%
Cyclohexane	C_6H_{12}	84.16	>99.9%
Methylcyclohexane	C_7H_{14}	98.19	>99.7%
Benzene	C_6H_6	78.11	>99.9%
Toluene	C_7H_8	92.14	>99.9%

Apparatus

Excess molar enthalpies have been measured with a quasi-isothermal flow calorimeter previously described [6]. The calorimeter is thermostatted at $T = (298.15 \pm 0.01)$ K or at $T = (313.15 \pm 0.01)$ K. The uncertainty in the measure of temperature is estimated to be less than 0.05 K. The H^E is calculated from differences in the heating power control, once the calibration procedure has been performed.

Knowing the volumetric flow rates delivered, the molar masses and the densities of the pure compounds, the mole fractions of the mixtures obtained in the mixing coil can be calculated. The maximum absolute uncertainty of mole fraction at equimolar composition is ±0.0008. Densities of pure liquids are determined by interpolating density data obtained from [12] at the measured temperature of delivery. Estimated densities at $T = 298.15$ K, were 0.76417, 0.80575, 0.67946, 0.68780, 0.66848, 0.77389, 0.76470, 0.87360 and 0.86219 g·cm^{-3} for the DBE, 1-butanol, heptane, 2,2,4-trimethylpentane, 1-hexene, cyclohexane, methylcyclohexane, benzene and toluene respectively. Mixtures of different compositions are studied to determine the dependence of H^E on mole fraction. The estimated relative uncertainty of the determined $H^E/\text{J}\cdot\text{mol}^{-1}$ is $\pm\, 0.01\cdot H^E$.

Excess Enthalpy Results

Concerning mixtures of liquids, thermodynamic excess properties are defined as the difference between an actual property value and the value calculated for an ideal solution at the same temperature, pressure and composition. A full explanation of thermodynamic properties of mixtures can be found in [13]. Then, for the molar excess enthalpy of a mixture, H^E, it stands as

$$H^E = H - H^{id} \tag{1}$$

For this property, its value corresponds to the enthalpy change of mixing, and then the ideal solution is represented by the linear contribution of the enthalpy of every pure compound present in the mixture,

$$H^E = H - \sum x_i H_i \tag{2}$$

being H_i the molar enthalpy of the pure compound and x_i the mole fraction of the respective compound in the mixture. For a binary mixture of compounds 1 and 2,

$$H^E = H - (x_1 H_1 + x_2 H_2) \tag{3}$$

Excess enthalpy provides direct information about the energetic effects arising between the molecules present in the mixture. The sign, magnitude, and symmetry of this quantity is a direct result of bond breaking and rearranging during the mixing process, and any effect arising from energetic interactions between both like and unlike molecules will be directly reflected in the enthalpy data and their representations. When the excess enthalpy is positive, mixing is endothermic while negative excess enthalpy signifies exothermic mixing. Intermolecular forces are the main constituent of changes in the enthalpy of a mixture. Stronger attractive forces between the mixed molecules, such as hydrogen-bonding, induced-dipole, and dipole-dipole interactions result in a lower enthalpy of the mixture and a release of heat. If strong interactions only exist between like-molecules the mixture will have a higher total enthalpy and absorb heat.

The experimental excess enthalpy of mixtures of DBE, or butanol, with heptane, iso-octane, 1-hexene, cyclohexane, methylcyclohexane, benzene, and toluene at 298.15 K and 313.15 K were previously measured by our group [6-11]

With respect to the mixtures of DBE + hydrocarbon, Figure 1 presents the behavior of the excess enthalpy versus composition at 298.15 K.

In mixtures of ether + hydrocarbon substances, which are non-polar substances, the dispersion forces are the most significant attractive molecular forces. The positive contribution to H^E associated with the disruption of interaction between like molecules with respect to the negative contribution due to the creation of interaction between unlike molecules explains the endothermic character ($H^E > 0$) of most of the mixtures of DBE + hydrocarbon, with the only exception of DBE + 1-hexene.

Concerning the mixtures of DBE + alkane, the behaviours of DBE + heptane and DBE + iso-octane at 298.15 K are very similar. For DBE + heptane the maximum value of the excess molar enthalpy is 123 J·mol^{-1}, obtained at an equimolar composition, while the maximum for DBE + iso-octane is 124 J·mol^{-1}, obtained at a mole fraction of DBE about 0.45.

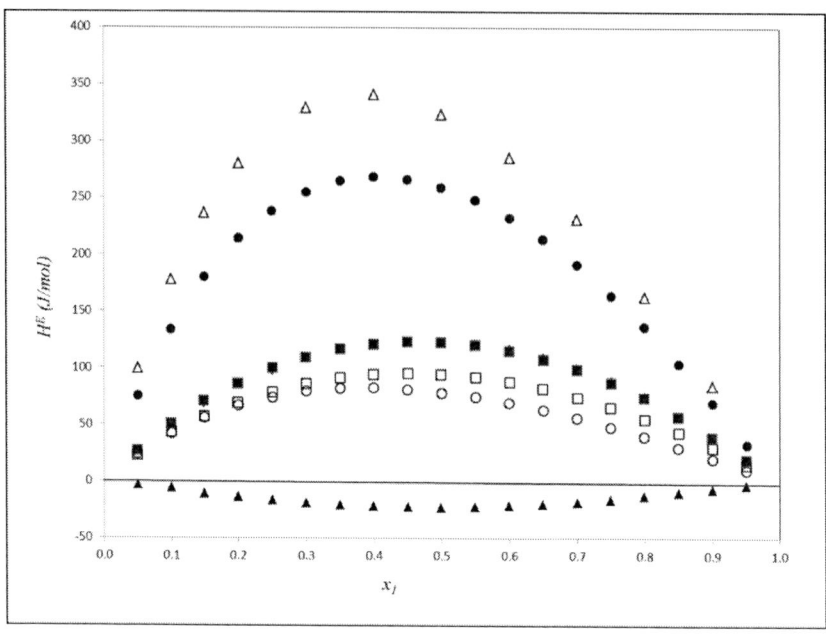

Figure 1. Excess molar enthalpy H^E at $T = 298.15$ K. Experimental results for DBE + heptane, (♦); DBE + 2, 2, 4 trimethylpentane (iso-octane), (■); DBE + 1-hexene, (▲); DBE + cyclohexane, (●); DBE + methylcyclohexane, (□); DBE + benzene, (Δ); DBE + toluene, (o).

Higher values of H^E are obtained for DBE + cyclohexane, with a maximum value of the molar excess enthalpy of 267 J·mol^{-1}, obtained at a mole fraction of DBE about 0.40. When replacing the cyclic alkane with the branched cycloalkane, methylcyclohexane, the maximum reaches only 96 J·mol^{-1}, showing the effect of the methyl group in the molecule. The same occurs when comparing the DBE + benzene (maximum 341 J·mol^{-1}, at a mole fraction of DBE about 0.40) with DBE + toluene (maximum 83 J·mol^{-1} at $x = 0.4$), reflecting the decreasing effect of the methyl group.

The excess molar enthalpy of the binary system DBE + 1-hexene at 298.15 K presents exothermic behavior ($H^E < 0$) in the whole range of composition, being the maximum -23 J·mol^{-1} at the equimolar composition.

The endothermic character of the mixtures of DBE + hydrocarbon reveals the positive contribution to H^E associated with the disruption of interaction between like molecules respect to the negative contribution due to the creation of interaction between unlike molecules. Our results show that for the given ether DBE, H^E (benzene) > H^E (cyclohexane) > H^E (heptane, 2,2,4, trimethylpentane > H^E (methylcyclohexane) > H^E (toluene) > H^E (1-hexene). Only the mixture DBE + 1-hexene shows exothermic effect in the entire range of composition at both temperatures, due to rather strong interactions between the double bond of 1-hexene and the oxygen of DBE which largely compensate for the endothermic effects due to both DBE and hexane upon mixing. The H^E curves are skewed towards small mole fractions of ether, reflecting the more active behavior of ether molecules.

Figure 2 shows the same for the mixtures of DBE + hydrocarbon at 313.15 K. Because of the increase in kinetic energy of molecules with the temperature, a decrease in the dispersion forces is expected and therefore, the endothermic effect is lower.

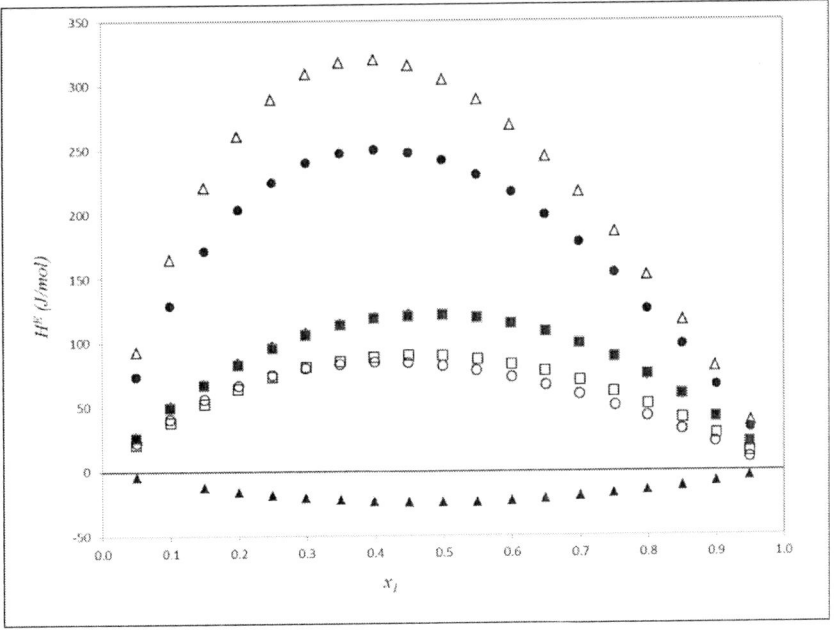

Figure 2. Excess molar enthalpy H^E at T = 313.15 K. Experimental results for DBE + heptane, (□); DBE + 2, 2, 4 trimethylpentane (iso-octane), (□); DBE + 1-hexene, (▲); DBE + cyclohexane, (●); DBE + methylcyclohexane, (□); DBE + benzene, (Δ); DBE + toluene, (o).

With respect to the binary mixtures of 1-butanol + hydrocarbon, H^E results are presented in Figures 3 and 4. All of the mixtures present endothermic behavior in the whole range of composition. The H^E curves are skewed towards low mole fractions of alcohol, reflecting their strong self-association character.

The maximum values of H^E at 298.15 K are the following for the mixtures of 1-butanol + hydrocarbon: (i) + heptane, 628 J·mol^{-1} at the mole fraction of 0.65 on heptane; (ii) + iso-octane, 636 J·mol^{-1}, at a mole fraction of TMP about 0.65; (iii) + 1-hexene, 648 J·mol^{-1} at the mole fraction of 0.7 on 1-hexene; (iv) + cyclohexane, 635 J·mol^{-1} at the mole fraction of 0.65 on cyclohexane; (v) + methyl cyclohexane, 588 J·mol^{-1} at the mole fraction of 0.65 of methylcyclohexane; (vi) + benzene, 1118 J·mol^{-1} at the mole fraction of 0.65 of benzene; (vii) + toluene, 1012 J·mol^{-1} at the mole fraction of 0.65 of toluene.

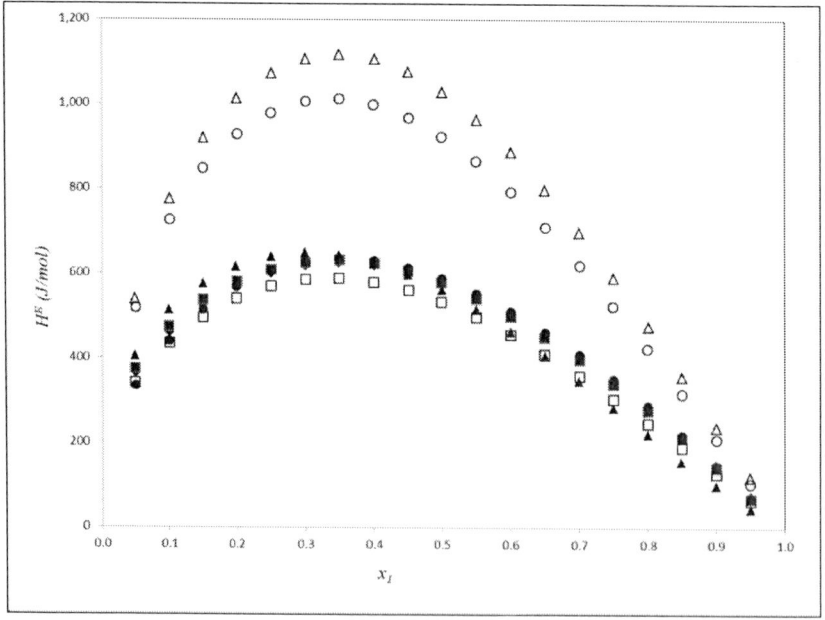

Figure 3. Excess molar enthalpy H^E at T = 298.15 K. Experimental results for 1-butanol + heptane, (□); 1-butanol + 2, 2, 4 trimethylpentane (iso-octane), (□); 1-butanol + 1-hexene, (▲); 1-butanol + cyclohexane, (●); 1-butanol + methylcyclohexane, (□); 1-butanol + benzene, (Δ); 1-butanol + toluene, (o).

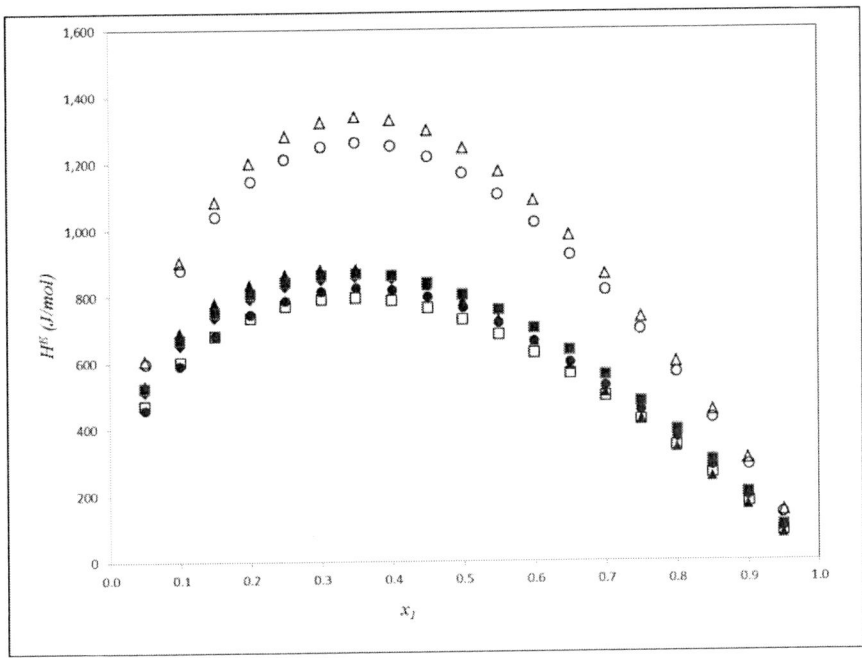

Figure 4. Excess molar enthalpy H^E at T = 313.15 K. Experimental results for 1-butanol + heptane, (☐); 1-butanol + 2, 2, 4 trimethylpentane (iso-octane), (☐); 1-butanol + 1-hexene, (▲); 1-butanol + cyclohexane, (●); 1-butanol + methylcyclohexane, (☐); 1-butanol + benzene, (Δ); 1-butanol + toluene, (o).

The chemical forces of the hydrogen bonds in the alkanol are stronger than the dispersion forces of the hydrocarbon and an endothermic character of the mixture is expected. This effect is enhanced by the increase of the temperature, and by the weakness of the dispersion forces. Experimental H^E data show that for the given 1-butanol, H^E (benzene) > H^E (toluene) > H^E (1-hexene) > H^E (heptane, 2,2,4, trimethylpentane) > H^E (cyclohexane) > H^E (methylcyclohexane).

An interesting case is the comparison of the interactions between DBE and benzene or toluene. The difference observed results from stronger interactions with toluene than with benzene, which compensates more the endothermic effects of disruption of molecular interactions between like molecules in toluene and in benzene respectively. Similarly, the less positive enthalpy for 1-butanol with toluene than with benzene results from stronger interactions with toluene than with benzene. As a matter of fact, the methyl group of toluene has an inductive effect for stronger interactions and at the same time reduces considerably the well know stacking organization of benzene molecules.

LITERATURE REVIEW

As an additional informational material, a literature search on excess enthalpy of liquid DBE and 1-butanol in mixture with the selected hydrocarbons has been obtained using the on-line version of the NIST scientific database ThermoLit[©]. The review includes only the interval of temperature and pressure of every property reported.

For interested readers, Table 2 reports 18 literature references related to excess enthalpy data for the mixtures of DBE with the selected hydrocarbons. All the references correspond to measurements at 101 kPa. The temperature interval of the reported data ranges from 298.15 K to 313.15 K. Most of the data are reported at one only temperature, being 298.15 K the most frequent.

Concerning the mixtures with 1-butanol, Table 3 reports 21 literature references related to excess enthalpy data of mixtures of 1-butanol + selected hydrocarbon. All the references correspond to measurements at 101 kPa, with the exception of reference [34], which reaches 400 kPa. This reference [34] is the only one that, at the same time, reports data up to 363.15 K, while the rest of references published data in the range from 283.15 K to 328.15 K. The same as for mixtures of DBE + hydrocarbon, most of the data are reported at 298.15 K.

From both tables, it can be observed that there are scarce data available for the mixtures of both oxygenated compounds with the alkene 1-hexene (which exists in fuels and bio-fuels but in a limited proportion) and methylcyclohexane. Meanwhile, mixtures with linear or branched alkanes, cycloalkanes and aromatics are more frequent, as they present more industrial applications.

Table 2. Reported excess enthalpy for binary mixtures of DBE + hydrocarbon

Reference	Year	T_{min}/K	T_{max}/K	P_{min}/kPa	P_{max}/kPa
Heptane					
[14]	1982	298.15	298.15	101	101
[15]	1988	298.15	298.15	101	101
[16]	1988	298.15	298.15	101	101
[17]	2000	298.15	298.15	101	101
[18]	2004	298.15	298.15	101	101
[9]	2010	298.15	313.15	101	101
2, 2, 4 Trimethylpentane					
[19]	2002	298.15	298.15	101	101
[20]	2008	298.15	298.15	101	101
[10]	2009	298.15	298.15	101	101
[11]	2012	313.15	313.15	101	101
1-Hexene					
[21]	2004	298.15	298.15	101	101
[11]	2012	298.15	313.15	101	101
Cyclohexane					
[22]	1985	298.15	298.15	101	101
[16]	1988	298.15	298.15	101	101
[23]	1991	303.15	303.15	101	101
[6]	2009	298.15	298.15	101	101
[11]	2012	313.15	313.15	101	101
Methylcyclohexane					
[11]	2012	298.15	313.15	101	101
Benzene					
[24]	1981	298.15	308.15	101	101
[25]	1982	298.15	298.15	101	101
[22]	1985	298.15	298.15	101	101
[24]	1991	303.15	303.15	101	101
[18]	2004	298.15	298.15	101	101
[8]	2009	298.15	313.15	101	101
Toluene					
[25]	1982	298.15	298.15	101	101
[22]	1985	298.15	298.15	101	101
[23]	1991	303.15	303.15	101	101
[18]	2004	298.15	298.15	101	101
[11]	2012	298.15	313.15	101	101

Table 3. Reported excess enthalpy for binary mixtures of 1-butanol + hydrocarbon

Reference	Year	T_{min}/K	T_{max}/K	P_{min}/kPa	P_{max}/kPa
Heptane					
[26]	1965	303.15	318.15	101	101
[27]	1975	288.15	328.15	101	101
[22]	1985	298.15	298.15	101	101
[28]	1986	298.15	298.15	101	101
[17]	2000	298.15	298.15	101	101
[9]	2010	298.15	313.15	101	101
2, 2, 4 Trimethylpentane					
[29]	1979	298.15	298.15	101	101
[10]	2010	298.15	298.15	101	101
[11]	2012	313.15	313.15	101	101
[30]	2012	298.15	313.15	101	101
1-Hexene					
[11]	2012	298.15	313.15	101	101
Cyclohexane					
[31]	1983	288.15	318.15	101	101
[32]	1985	298.15	298.15	101	101
[33]	1986	298.15	298.15	101	101
[34]	1995	283.15	363.15	400	400
[6]	2009	298.15	298.15	101	101
[11]	2012	313.15	313.15	101	101
Methylcyclohexane					
[35]	1987	323.15	323.15	101	101
[11]	2012	298.15	313.15	101	101
Benzene					
[36]	1961	298.15	318.15	101	101
[37]	1961	298.15	318.15	101	101
[33]	1986	298.15	298.15	101	101
[38]	1988	288.15	323.15	101	101
[8]	2009	298.15	313.15	101	101
[39]	2010	303.15	303.15	101	101
Toluene					
[36]	1961	298.15	318.15	101	101
[11]	2012	298.15	313.15	101	101

CONCLUSION

Isothermal excess molar enthalpies at $T = 298.15$ K and $T = 313.15$ K for the binary systems DBE or 1-butanol and heptane, or 2,2,4-trimethylpentane (iso-octane), or 1-hexene, or cyclohexane, or methylcyclohexane, or benzene, or toluene, at atmospheric pressure, measured by using an isothermal flow calorimeter, were compared. All of the binary systems DBE + hydrocarbon show endothermic effect and slight asymmetric H^E behavior at the measured temperatures. Only the binary mixture DBE + 1-hexene, at 298.15 K and 313.15 K, shows exothermic behavior. All of the 1-butanol + hydrocarbon mixtures show endothermic effect and strong asymmetric H^E behavior at the measured temperatures. Intermolecular and association effects involved in these systems have been discussed and presented.

In addition, a literature search on existing excess enthalpy of mixtures of DBE or 1-butanol with the selected hydrocarbons data is presented. Data at a pressure above the atmospheric date 101 kPa are almost non-existent, with the exception of one reference for 1-butanol + cyclohexane up to 400 kPa. Concerning the interval of temperature, the date reported are within a moderate range, from 283.15 K to 328.15 K, with the same exception of the mixture 1-butanol + cyclohexane, which was measured up to 363.15 kPa.

REFERENCES

[1] *Directive 2009/28/EC of the European Parliament and of the Council on the promotion of the use of energy from renewable sources.*

[2] *Directive 2009/30/EC of the European Parliament and of the Council as regards the specification of petrol, diesel and gas-oil and introducing a mechanism to monitor and reduce greenhouse gas emissions.*

[3] Ezeji, T. C., Blaschek, H. P. 2010. "Butanol Production from Lignocellulosic Biomass". In *Biofuels from Agricultural Wastes and Byproducts*, edited by H. P. Blaschek, T. C. Ezeji and J. Sheffran. Ames: Wiley-Blackwell.

[4] Agarwal A. K. (2007). Biofuels (alcohols and biodiesel) applications as fuels for internal combustion engines. *Progress in Energy and Combustion Science*, 33: 233-271.

[5] Surisetty, V. R., Dalai A. K., Kozinski J. (2011). Alcohols as alternative fuels: An overview. *Applied Catalysis A: General*, 404: 1–11.

[6] Aguilar, F., Alaoui, F. E. M., Alonso-Tristán, C., Segovia, J. J., Villamañán, M. A., Montero, E. A. (2009). Excess Enthalpies of Binary and Ternary Mixtures Containing Dibutyl Ether, Cyclohexane, and 1-Butanol at 298.15 K. *J. Chem. Eng. Data*, 54: 1672-1679.

[7] Aguilar, F., Alaoui, F. E. M., Alonso-Tristán, C., Segovia, J. J., Villamañán, M. A., Montero, E. A. (2009). Excess Enthalpies of Binary and Ternary Mixtures Containing Dibutyl Ether, Cyclohexane, and 1-Butanol at 298.15 K. Corrections. *J. Chem. Eng. Data*, 54: 2341-2342.

[8] Aguilar, F., Alaoui, F. E. M., Segovia, J. J., Villamañán, M. A., Montero, E. A. (2009). Excess enthalpies of ether + alcohol + hydrocarbon mixtures: Binary and ternary mixtures containing dibutyl ether (DBE), 1-butanol and benzene at 298.15K and 313.15 K. *Fluid Phase Equilib.*, 284: 106-113.

[9] Aguilar, F., Alaoui, F. E. M., Segovia, J. J., Villamañán, M. A., Montero, E. A. (2010). Excess enthalpies of binary and ternary mixtures containing dibutyl ether (DBE), 1-butanol, and heptane at T = 298.15 K and 313.15 K. *J. Chem. Thermodyn.*, 42: 28-37.

[10] Aguilar, F., Alaoui, F. E. M., Segovia, J. J., Villamañán, M. A., Montero, E. A. (2010). Excess enthalpies of oxygenated compounds + hydrocarbon mixtures: binary and ternary mixtures containing dibutyl ether (DBE), 1-butanol and 2, 2, 4 trimethylpentane at 298.15 K. *Fluid Phase Equilib.*, 290: 15-20.

[11] Aguilar, F., Alaoui, F. E. M., Segovia, J. J., Villamañán, M. A., Montero, E. A. (2012). Ether + alcohol + hydrocarbon mixtures in fuels and bio-fuels: excess enthalpies of binary mixtures containing dibutyl ether (DBE) or 1-butanol and 1hexene or methylcyclohexane or toluene or cyclohexane or 2, 2, 4 trimethylpentane at 298.15 K and 313.15 K. *Fluid Phase Equilib.*, 315: 1-8.

[12] Riddick, J. A., Bunger, W. B., Sakano, T. K. 1986. *Organic Solvents. Physical Properties and Methods of Purification.* Wiley; New York.

[13] Van Ness, H. C., Abbott, M. M. 1982. *Classical Thermodynamics of Nonelectrolite Solutions with Applications to Phase Equilibria.* McGraw-Hill; New York.

[14] Villamañán, M. A., Casanova, C., Roux, A. H., Grolier, J.-P. E. (1982). Excess enthalpies of some binary mixtures. n-alkane + aliphatic ether, n- alkane + hydroxy ether, aliphatic ether + hydroxy ether. *J. Chem. Eng. Data,* 27: 89-91.

[15] Benson, G. C., Luo, B., Lu, B. C.-Y. (1988). Excess enthalpies of dibutyl ether + n-alkane mixtures at 298.15 K. *Can. J. Chem.,* 66: 531-534.

[16] Marongiu, B., Dernini, S., Lepori, L., Matteoli, E., Kehiaian, H. V. (1988). Thermodynamics of Binary Mixtures Containing Ethers of Acetals 1. Excess Enthalpies of Linear Ethers or Acetals + Heptane of + Cyclohexane Mixtures. *J. Chem. Eng. Data,* 33: 118-122.

[17] Rezanova, E. N., Kammerer, K., Lichtenthaler, R. N. (2000). Excess enthalpies and volumes of ternary mixtures containing 1-propanol or 1-butanol, and ether (diisopropyl ether or dibutyl ether), and heptane. *J. Chem. Eng. Data,* 45: 124-130.

[18] Bernazzani, L., Gianni, P., Mollica, V., Pizzolla, P. (2004). Group contributions to enthalpies of solvation in octan-1-ol and di-n-butyl ether. *Thermochim. Acta,* 418: 109-116.

[19] Peng, D.-Y., Benson, G. C., Lu, B. C.-Y. (2002). Excess enthalpies of (di-n-butyl ether + 2,2,4-trimethylpentane + heptane, or octane) at the temperature 298.15 K. *J. Chem. Thermodyn.,* 34: 413-422.

[20] Kim, M.-G., Park, S.-J., Hwang, I.-C. (2008). Excess molar enthalpies for the binary and ternary mixtures of ether compounds (di-isopropyl ether, di-butyl ether, propyl vinyl ether) with ethanol and isooctane at 298.15K. *Korean J. Chem. Eng.*, 25: 1160-1164.

[21] Wang, Z., Benson G. C., Lu, B. C.-Y. (2004). Excess Enthalpies of Binary Mixtures of 1-Hexene with Some Ethers at 25+-C. *J. Solution Chem.*, 33: 143-147.

[22] Stephenson, W. K., Fuchs, R. (1985). Enthalpies of interaction of aromatic solutes with organic solvents. *Can. J. Chem.*, 663: 2529-2534.

[23] Sharma, S. C., Kumar, P., Syngal, M. (1991). Excess enthalpies of (di-n-butyl ether + cyclohexane or benzene or toluene or ethyl benzene or m-xylene or p-xylene or mesitylene) at the temperature 303.15 K. *J. Chem. Thermodyn.*, 23: 43-47.

[24] Ott, J. B., Marsh, K. N., Richards, A. E. (1981). Excess enthalpies, excess Gibbs free energies and excess volumes for di-n- butyl ether + benzene and excess Gibbs free energies and excess volumes for di-n-butyl ether + tetrachloromethane at 298.15 K. *J. Chem. Thermodyn.*, 13: 447-455.

[25] Fuchs, R., Peacock, L. A., Stephenson, W. K. (1982). Enthalpies of interaction of polar and nonpolar molecules with aromatic solvents. *Can. J. Chem.*, 60: 1953-1958.

[26] Savini, C. G., Winterhalter, D. R., Van Ness, H. C. (1965). Heats of Mixing of Some Alcohols-Hydrocarbon Systems. *J. Chem. Eng. Data*, 10: 168-171.

[27] Nguyen, T. H., Ratcliff, G. A. (1975). Heats of Mixing of n-Alcohol-n-Alkane Systems at 15°C and 55°C. *J. Chem. Eng. Data*, 20: 252-255.

[28] Oswald, G., Schmittecker, B., Wagner, D., Lichtenthaler, R. N. (1986). Excess enthalpies and excess volumes of alkanol + n-heptane mixtures at high pressures. *Fluid Phase Equilib.*, 27: 119-135.

[29] Stokes, R. H., Adamson, M., Richards, A. (1979). Excess enthalpies of 2,2,4-trimethylpentane with several alcohols at low mole fractions of 2,2,4-trimethylpentane. *J. Chem. Thermodyn.*, 11: 303-304.

[30] Faneite, A. M., Garces, S. I., Aular, J. A., Urdaneta, M. R., Soto, D. (2012). Excess molar volumes, excess molar enthalpies and refractive index deviations for binary mixtures of propan-1-ol, butan-1-ol and pentan-1-ol with 2,2,4-trimethylpentane at 298.15 K. *Fluid Phase Equilib.,* 334: 117-127.

[31] French, H. T. (1983). Thermodynamic Functions of the Systems 1-Butanol, 2-Butanol, and t-Butanol + Cyclohexane. *J. Solution Chem.*, 12: 869-887.

[32] Stephenson, W. K., Fuchs, R. (1985). Enthalpies of interaction of hydroxylic solutes with organic solvent. *Can. J. Chem.*, 63: 2535-2539.

[33] Saris, P., Rosenholm, J. B., Sjoblom, E., Henriksson, U. (1986). A thermometric investigation of the association equilibria of alcohols in hydrocarbons. *J. Phys. Chem.,* 90: 660-665.

[34] Lowen, B., Schulz, S. (1995). Excess molar enthalpies of cyclohexane + n-alcohols at 283.25, 298.15, 323.15, 343.15 and 363.15 K and at a pressure of 0.4 MPa. *Thermochim. Acta,* 265: 63-71.

[35] Alonso, R., Guerrero, R., Corrales, J. A. (1987). Excess Molar Enthalpies of (Methylcyclohexane + an Alcohol) at 323.15 K I. Results for n-Butanol, n-Pentanol, and n-Hexanol. *J. Chem. Thermodyn.,* 19: 1271-1273.

[36] Mrazek, R. V., Van Ness, H. C. (1961). Heats of mixing: alcohol - aromatic binary systems at 25, 35 and 45°C. *AIChE J.,* 7: 190-195.

[37] Brown, I., Fock, W. (1961). Heats of mixing IV. Solutions of n-alcohols with benzene at 25, 35, and 45°C. *Aust. J. Chem.*, 1961, 14, 387-396.

[38] Chao, J. P., Dai, M. (1988). Studies on the thermodynamic properties of binary systems containing alcohols.VII.Temperature dependence on excess enthalpies for n-propanol + benzene and n-butanol+benzene. *Thermochim. Acta,* 123: 285-291.

[39] Didaoui, S., Ait-Kaci, A. (2010). Experimental and predicted excess molar enthalpies of binary mixtures containing 2,2'-oxybis[propane],

benzene, butan-1-ol, 2-methylpropan-1-ol, 2-methyl-2-ene-1-propanol at 303.15 K. *Fluid Phase Equilib.*, 299: 60-64.

BIOGRAPHICAL SKETCHES

Eduardo Montero

Affiliation: Universidad de Burgos, Spain

Education: PhD (1996), MSc Industrial Engineering (1984)

Research and Professional Experience: He has been employed in the Department of Electromechanical Engineering at the University of Burgos, Spain, since 1984, being Assistant Professor since 1988 and Full Professor since 2017.

His academic expertise includes engineering thermodynamics, heat transfer and energy technology. His current research interests are experimental determination of thermodynamic properties of fluid mixtures (biofuels, refrigerants), energy efficiency systems in buildings and thermal energy storage. For more than 30 years, he has been the Director of research of the Energy Engineering group at the University of Burgos, where he has developed international and national competitive research projects, as well as research services for industry. He is co-author of more than 60 articles published in high impact factor journals in the field of energy and fluids.

Professional Appointments: At the University of Burgos he has hold the positions of Vice-Dean of External Relations at the Escuela Politécnica Superior (1989-1994), Vice-Rector of Economics and Planning (1994-1997) and Head of the Department of Electromechanical Engineering (2004-2012).

Fernando Aguilar

Affiliation: Universidad de Burgos, Spain

Education: PhD (2010), MSc Industrial Engineering (2000), BSc Mechanical Engineering (1989)

Research and Professional Experience: He has been employed in the Department of Electromechanical Engineering at the University of Burgos, Spain, holding the position of Assistant Professor since 1994

His academic expertise includes engineering thermodynamics, heat transfer and heat engines. His current research interests are new renewable fluids (biofuels, refrigerants), energy efficiency systems in buildings and thermal energy storage. For more than 25 years, he has been member of the Energy Engineering group at the University of Burgos, where he has participated in international and national competitive research projects, as well as research services for industry. He is co-author of more than 45 articles published in high impact factor journals in the field of energy and fluids.

Fatima Ezzahrae M'Hamdi Alaoui

Affiliation: Université Chouaïb Doukkali d'El Jadida-Maroc.

Education: PhD (2011), MSc Engineering Thermodynamics of Fluids (2009)

Research and Professional Experience: She has been employed in the Département Sciences et Technologies Industrielles at the Université Chouaïb Doukkali d'El Jadida, Morocco, holding the position of Assistant Professor since 2013.

Her academic expertise includes engineering thermodynamics and environmental sciences. Her current research interests are new renewable fluids and biomass. She belongs to the Laboratoire des Sciences de l'Ingénieur Pour l'Energie at the Université Chouaïb Doukkali d'El Jadida, where she participates in international and national competitive research projects. She is co-author of more than 28 articles published in high impact factor journals in the field of energy and fluids.

Natalia Muñoz-Rujas

Affiliation: Universidad de Burgos, Spain

Education: PhD (2018), MSc Engineering Thermodynamics (2012)

Research and Professional Experience: She has been employed in the Department of Electromechanical Engineering at the University of Burgos, Spain, holding the position of Lecturer since 2018.

Her academic expertise includes engineering thermodynamics and fluid mechanics. His current research interests are new renewable fluids (biofuels, refrigerants) and thermal energy storage. She has recently joined the Energy Engineering group at the University of Burgos. She is co-author of 16 articles published in high impact factor journals in the field of energy and fluids.

Publications from the Last 3 Years:

Abala, I., Alaoui, F. E. M., Eddine, A. S., Aguilar, F., Muñoz-Rujas, N., Montero, E. (2019). (ρ, VE, T) Measurements of the Ternary Mixture (Dibutyl Ether + 1-Heptanol + Heptane) at Temperatures up to 393.15 K and Pressures up to 140 MPa and Modeling Using the Peng–Robinson and PC-SAFT Equations of State. *J. Chem. Eng. Data*, 64: 3861-3873.

Abala, I., M'hamdi Alaoui, F. E., Chhiti, Y., Eddine, A. S., Muñoz-Rujas, N., Aguilar, F. (2019). Density of biofuel mixtures (Dibutyl ether + Heptane) at temperatures from (298.15–393.15) K and at pressures up to 140 MPa: Experimental data and PC-SAFT modeling. *Fluid Phase Equilib.*, 491: 35-44.

Abala, I., M'hamdi Alaoui, F. E., Chhiti, Y., Eddine, A. S., Muñoz-Rujas, N., Aguilar, F. (2019). Experimental density and PC-SAFT modeling of biofuel mixtures (DBE + 1-Heptanol) at temperatures from (298.15 to 393.15) K and at pressures up to 140 MPa. *J. Chem. Thermodyn.*, 131: 269-285.

Aitbelale, R., Abala, I., M'hamdi Alaoui, F. E., Eddine, A. S., Muñoz-Rujas, N., Aguilar, F. (2019). Characterization and determination of thermodynamic properties of waste cooking oil biodiesel: Experimental, correlation and modeling density over a wide temperature range up to 393.15 and pressure up to 140 MPa. *Fluid Phase Equilib.*, 497: 87-96.

Aitbelale, R., Chhiti, Y., M'hamdi Alaoui, F. E., Eddine, A. S., Muñoz-Rujas, N., Aguilar, F. (2019). High-Pressure Soybean Oil Biodiesel Density: Experimental Measurements, Correlation by Tait Equation, and Perturbed Chain SAFT (PC-SAFT) Modeling". *J. Chem. Eng. Data*, 64: 3994-4004.

Alaoui, F. E. M., Aguilar, F., González-Fernández, M. J., Montero, E. A. (2017). Excess enthalpies of ternary mixtures of (oxygenated additives + cycloalkane) in fuels and bio-fuels: (dibutyl ether + 1-propanol + cyclohexane), or methylcyclohexane, at T = (298.15 and 313.15) K. *J. Chem. Thermodyn.*, 105: 112-122.

Boumanchar, I., Chhiti, Y., M'hamdi Alaoui, F. E., Elkhouakni, M., Dine, A. S., Bentiss, F., Jama, C. (2019). Investigation of (co)-combustion kinetics of biomass, coal and municipal solid wastes. *Waste Management*, 97: 10-18.

Boumanchar, I., Chhiti, Y., M'hamdi Alaoui, F. E., El-Ouinani, A., Dine, A. S., Bentiss, F., Jama, C., Bensitel, M. (2017). Effect of materials mixture on the higher heating value: Case of biomass, biochar and municipal solid waste. *Waste Management* 61: 78-86.

Briones-Llorente, R., Calderón, V., Gutiérrez-González, S., Montero, E., Rodríguez, (2019). Testing of the Integrated Energy Behavior of

Sustainable Improved Mortar Panels with Recycled Additives by Means of Energy Simulation. *Sustainability*, 11: 3117.

Dakkach, M., Aguilar, F., Alaoui, F. E. M., Montero, E. A. (2017). Liquid densities and excess volumes of biofuel mixtures: (2-butanol + di-isopropyl ether) system at pressures up to 140 MPa and temperatures from 293.15 K to 393.28 K. *J. Chem. Thermodyn.*, 105: 123-132.

Dakkach, M., Muñoz-Rujas, N., Aguilar, F., Alaoui, F. E. M., Montero, E. A. (2018). High pressure and high temperature volumetric properties of (2-propanol and di-isopropyl ether) system. *Fluid Phase Equilib.*, 469: 33-39.

Darkaoui, M., Muñoz-Rujas, N., Aguilar, F., El Amarti, A., Dakkach, M., Montero, E. A. (2018). Liquid Density of Mixtures of Methyl Nonafluorobutyl Ether (HFE-7100) + n-Heptane at Pressures up to 80 MPa and Temperatures from 298.15 to 393.15 K, *J. Chem. Eng. Data*, 63: 2966–2974.

Makhlouf, H., Muñoz-Rujas, N., Aguilar, F., Belhachemi, B., Montero, E. A., Bahadur, I., Negadi, L. (2019). Density, speed of sound and refractive index of mixtures containing 2-phenoxyethanol with propanol or butanol at various temperatures, *J. Chem. Thermodyn.*, 128: 394-405.

Montero, E. A., Aguilar, F., Muñoz-Rujas, N., Alaoui, F. E. M. 2017. "Thermodynamic properties of propanol and butanol as oxygenate additives to biofuels". In *Frontiers in Bioenergy and Biofuels*, edited by Eduardo Jacob-Lopes and Leila Queiroz Zepka, InTechOpen, Rijeka (Croatia), ISBN 978-953-51-2892-2, Print ISBN 978-953-51-2891-5. DOI: 10.5772/66297

Muñoz-Rujas, N., Aguilar, F., García-Alonso, J. M., Montero, E. A. (2019). Thermodynamics of binary mixtures 1-ethoxy-1,1,2,2,3,3,4,4,4-nonafluorobutane (HFE-7200) + 2-propanol: High pressure density, speed of sound and derivative properties. *J. Chem. Thermodyn.*, 131: 630-647.

Muñoz-Rujas, N., Aguilar, F., García-Alonso, J. M., Montero, E. A. (2018). High pressure density and speed of sound of hydrofluoroether fluid 1,1,1,2,2,3,4,5,5,5-decafluoro-3-methoxy-4-(trifluoromethyl)-pentane (HFE-7300). *J. Chem. Thermodyn.,* 121: 1-7.

Muñoz-Rujas, N., Bazile, J. P., Aguilar, F., Galliero, G., Montero, E., Daridon, J. L. (2019). Speed of sound, density and derivative properties of binary mixtures HFE-7500 + Diisopropyl ether under high pressure, *J. Chem. Thermodyn.,* 128,: 19-33.

Muñoz-Rujas, N., Bazile, J. P., Aguilar, F., Galliero, G., Montero, E., Daridon, J. L. (2017). Speed of sound and derivative properties of diisopropyl ether under high pressure. *Fluid Phase Equilib.,* 449: 148-155.

Muñoz-Rujas, N., Bazile, J. P., Aguilar, F., Galliero, G., Montero, E., Daridon, J. L. (2017). Speed of sound and derivative properties of hydrofluoroether fluid HFE-7500 under high pressure. *J. Chem. Thermodyn.,* 112: 52-58.

Rubio-Pérez, G., Muñoz-Rujas, N., Srhiyer, A., Montero, E. A., Aguilar, F. (2018). Isobaric vapor-liquid equilibrium, density and speed of sound of binary mixtures 2,2,4-trimethylpentane and 1-butanol or dibutyl ether (DBE) at 101.3 kPa. *Fluid Phase Equilib.,* 475: 10-17

Srhiyer, A., Muñoz-Rujas, N., Aguilar, F., Segovia, J. J., Montero, E. A. (2017). High pressure volumetric properties of the binary mixtures di-isopropyl ether + 2,2,4-trimethylpentane. *J. Chem. Eng. Data,* 62: 3610-3619.

Srhiyer, A., Muñoz-Rujas, N., Aguilar, F., Segovia, J. J., Montero, E. A. (2017). High pressure liquid densities and excess volumes of the (di-isopropyl ether + 1-hexanol) system. *J. Chem. Thermodyn.,* 113: 213-218.

In: Biofuels
Editor: George R. Carey

ISBN: 978-1-53617-721-3
© 2020 Nova Science Publishers, Inc.

Chapter 3

ENVIRONMENTAL ASPECTS OF USING BIODIESEL AS A SUSTAINABLE ENERGY SOURCE: CURRENT SITUATION AND FUTURE TRENDS

Mehdi Ardjmand[1], Farid Jafarihaghighi[1], Mehrdad Mirzajanzadeh[2], Aida Gifani[1] and Hasanali Bahrami[3]

[1]Chemical Engineering Department, South Tehran Branch, Islamic Azad University, Tehran, Iran
[2]Department of Chemical Engineering, Science & Research Branch, Islamic Azad University, Tehran, Iran
[3]Department of Mechatronics, Arak University, Iran

ABSTRACT

The current energy crisis in the era of increasing energy consumption, together with the increment in greenhouse gas concentrations (for instance: carbon dioxide, methane, and nitrous oxide, etc.) from burning petroleum-based fuels causing environmental issues, have made scientist consider substituting fossil fuels with renewable and clean fuels. Hence, great studies have been conducted regarding finding alternative fuels. Amongst fuels like methanol, liquefied petroleum gas (LPG), compressed natural gas (CNG), vegetable oils, liquefied natural gas (LNG), reformulated diesel fuel, and reformulated gasoline, vegetable oils are the only non–fossil fuels. Biodiesel is one of the promising alternatives derived from renewable resources, like animal fats or vegetable oils, which have been realized as an environmentally benign fuel because of not only its green advantages (sustainability, biodegradability, and non-toxicity), but also several extra social benefits, like creation of new jobs, less emission of soot and carbon, and less global warming. In this regard, an attempt has been made to briefly introduce the production of this fuel from different edible and non-edible feedstock to mention various types of homogeneous and heterogeneous acid or base catalyst applied for reactions. The benefits and drawbacks of biodiesel compared with diesel fuel are also included. Some of the advantages of biodiesel are 1-carbon cycle, 2- less emission, 3- improvement of combustion and reduction of emissions of unburned hydrocarbons due to oxygen presence, 4- higher lubricity, 5- higher flash point and higher cetane number, and etc.

Keywords: biodiesel, environment effects, sustainable energy

INTRODUCTION

Petroleum fuels play a very significant role in the growth of industrial progress, transportation, and agriculture (Choi and Oh, 2006; Najafi, 2018). In other words, development of societies has resulted in increased energy demand; particularly, after the Industrial Revolution. This increased energy demand has been mostly met through the combustion of various materials, such as oil, coal and natural gas. These resources are natural sources or fossil fuels, thus, non-renewable. High emissions of, NOx, CO_2, SO_2, particulate matter, poly-aromatic hydrocarbons, and hydrocarbons are the result of fossil fuel consumptions, creating environmental problems (Canakci and Van Gerpen, 2001; Ashraful et al., 2016). Statistically, fossil fuels account for about 88% of primary energy consumption (35% oil, 29% coal, and 24% natural gas). In this regard, daily reduction of finite fossil resources and deterioration of environmental concerns have made researchers consider alternative sources that have received much attention in recent years. Development of alternative energy options has been discussed as a contributor of sustainable economic growth in human societies (Rajak, Nashine, and Verma, 2019).

Numerous proposals have been made concerning the practicality and availability of an environmentally sound fuel that can be domestically provided. Methanol, compressed natural gas (CNG), ethanol, liquefied natural gas (LNG), liquefied petroleum gas (LPG), reformulated gasoline, reformulated diesel fuel, and vegetable oils have all been proposed as alternative fuels. Among these alternative fuels, vegetable oils and ethanol are the only non–fossil fuels (Hoseini et al., 2018).

As the formation of fossil fuels takes a long time, their availability can be prolonged over their consumption (Sharma and Singh, 2009). Accordingly, there is a dire need of finding a substitute to meet the energy demand of the world. Biodiesel that is derived from vegetable oils is a proper substitute of petroleum-based diesel due to its renewability and biodegradability. Biodiesel is the greatest candidate for diesel fuel replacement due to its applicability in any compression-ignition engine without anu requirement for modification. Furthermore, biodiesel does not comprise any aromatic hydrocarbons, metals, sulfur or crude oil residues since it is completely prepared through vegetable sources (Velmurugan et al., 2019; Saravankumar et al., 2019).

Compared to conventional diesel fuels, biodiesel is inclined to have less carbon monoxide and soot emissions. Unlike fossil fuels, the use of biodiesel does not affect the global warming, as the emitted CO_2 is once again absorbed by the plants grown for vegetable oil/biodiesel production. Running the engine with biodiesel can extend the diesel engine life since it is more lubricating than petroleum diesel fuel. These are some of the benefits of biodiesel in comparison to conventional fossil fuels; however, it has still some drawbacks (Pinto et al., 2005; Beer et al., 2001).

Different types of vegetable oils, like soybean or rapeseed, can be applied for biodiesel production. Lots of studies have employing vegetable oil, both in its modified and neat forms, for its production. Researches have revealed that the use of vegetable oil in its neat form is possible, but not preferable. High viscosity of vegetable oils and their low volatility affect atomization and spray pattern of fuel, resulting in incomplete combustion and severe carbon deposits, injector choking, and piston ring sticking (Bari, Yu, and Lim, 2002; Leung, Wu, and Leung, 2010).

Renewable energy use is one of the most efficient ways to increase its share in the world matrix. Most of the "new renewable energy sources" are still experiencing large-scale commercial growth; nevertheless, some technologies are already well established.

RENEWABLE SOURCE OF ENERGY

Each generation is confronted with new challenges and new opportunities and the the most fundamental issue of the 21st century is energy supply. Energy improves people's lives and standard of living. Food, water, health, environment, education, population, and war are strictly dependent on the availability of energy. Although the energy crisis in this era is a challenge, it opens up new windows and opportunities. It offers scientists a precious chance to take an active role in finding new solutions and improving energy technologies. Thinking about the energy constrains is what makes us protect the earth and be more concerned about our living world (Armaroli and Balzani, 2007).

In order to reduce greenhouse gases released into the atmosphere, renewable energy technologies are introduced in the energy generation market and have achieved sustainable development (Pinto et al., 2005; Rounce et al., 2009; Knothe, Krahl, and Van Gerpen, 2015).

Nowadays, renewable energies are fairly well distributed all over the world. Such a generalized increase in all fields of renewable energies is very important since it allows each country to construct a diversified energy portfolio. Solar energy, wind power, biodiesel, ethanol and geothermal energy (related to the heat stored in the depths of cooling planet) are such important renewable energies. Other minor energy resources can be obtained by taking advantage of ocean and lake temperature gradients, currents, and waves, and the earth-moon gravitational energy that can be exploited in some selected coastal areas of the northern hemisphere, where tides move large volumes of seawater in the relatively short time. All these kinds of renewable energies can compete with conventional source of energy (Nalgundwar, Paul, and Sharma, 2016; Armaroli and Balzani, 2007).

Solar energy is an abundant and economical energy source, known as clean energy that should be converted into useful forms of energy, like electrical power and thermal power. Its conversion to fuel is more advisable because of the intermittent nature of solar energy. Currently, fossil-fuel exploitation is an important challenge faced by human beings due to the progressive reduction of fossil fuel reserves, their harmful effects, and global warming that threatens human health (Armaroli and Balzani, 2007; Knothe, Krahl, and Van Gerpen, 2015).

The wind power industry produces electricity through wind turbines. According to the European Wind Energy Association, the vast European wind-energy potential will be able to meet the electricity needs in Europe (Canakci and Van Gerpen, 2001). Wind turbines, run-of-the-river hydroelectric plants, tidal and wave stations are all well-suited to be integrated into the electricity grid. In the transportation sector, the most effective method for getting wind energy to the road is a battery-electric fuel chain. But even these renewable energy resources still have environmental impacts. For instance, wind energy can lead to regional climate alterations (Mazza and Hammerschlag, 2005).

BIODIESEL

Biodiesel is a substitute for diesel fuel in CI engines and is described as long-chain fatty acid methyl/ethyl esters that is produced from renewable sources, such as vegetable oils (soybean oil, cottonseed oil, canola oil, and rapeseed oil), recycled cooking greases or oils (e.g., yellow grease), or animal fats (beef tallow and pork lard), and can be applied in compression ignition engines without any substantial changes (Buyukkaya, 2010). Renewable oils are obtained from plants that use sunlight and air to produce oil and can do so year after year on the cropland. Animal fats are also obtained when the animal consumes animals or plants, and these too are renewable. Cooking oils are frequently plant-based, but may comprise animal fats that are both renewable and recycled. In this regard, another difference between diesel and biodiesel is the presence of oxygen in the chemical structure of biodiesel, which decreases emissions of unburned hydrocarbons. Another advantage of biodiesel is that it is carbon neutral, meaning that it does not produce pure carbon. In other words, when the plants grow, they get the same amount of carbon dioxide produced by consuming biodiesel in the engines (Ashraful et al., 2016).

As was mentioned previously, direct use of biodiesel is possible in diesel engines, with just minor change of some simple pieces of the engine, like gaskets or filters. However, in some cases, loss of engine power, cold start problems, and material compatibility may be observed. Therefore, it is recommended to blend it with diesel as B5 or B20 (B20 composed of 80% diesel and 20% biodiesel). With these low ratios, the emission of greenhouse gases is prevented, even with no changes or modification of engine parts (Uyumaz, 2018). Biodiesel can reduce many harmful exhaust gases released from diesel vehicles. It has been observed that direct use of biodiesel reduces the emission of CO by 46.7%, particle matters by 66.7%, and HC by 45.2%. Currently, many countries, including the United States of America, Brazil, Germany, Australia, and Austria, use bioethanol in gasoline engines and a compound of biodiesel and diesel in diesel engines (Ma and Hanna, 1999).

There are many methods, such as direct use, microemulsion, thermal cracking, and transesterification, for biodiesel production and consumption. Direct use of biodiesel is not practical due to its high viscosity that can damage engine parts (Abdullah et al., 2019). Biodiesel obtained from microemulsion or thermal cracking methods is not suitable, leading to incomplete combustion of fuel due to lower cetane number. Transesterification is a very simple and conventional method for producing biodiesel. Many studies have been done on this method, utilizing it to convert oils to biodiesel in an industrial scale. In general, transesterification reaction can convert vegetable oils to alkyl esters or biodiesel to lower the viscosity of these oils and have similar properties to petroleum-based fuels (Ahmad et al., 2011).

Transesterification is a chemical reaction, where the ester is transformed into a different ester. Principally, the process consists of replacing the alkoxy group of an ester with that of an alcohol. This reaction is normally catalyzed by bases and acids. Generally, methanol is the alcohol predominantly used for biodiesel synthesis since it is the least expensive alcohol, although there are exceptions in some countries. Roughly speaking, producing 100 pounds of biodiesel requires 100 pounds of fat or oil with 10 pounds of a short-chain alcohol (usually methanol) in the presence of a catalyst (usually potassium hydroxide or sodium hydroxide). In this case, 10 pounds of glycerin will also be produced (Handling and Edition, 2009).

The reaction temperature and time depend on the type of catalyst and type of processes. However, in general, the reaction temperature is 55–70 °C and the reaction time is 1–2 h for conventional processes and catalysts. Later after the reaction takes place, the Fatty Acid Methyl Esters (FAME) and glycerol, as the products of the transesterification reaction, are simply separated with the help of gravity due to the difference in density and their insolubility in each other. The employment of membrane technology for the separation of transesterification reaction products or the production of biodiesel by membrane reactor has progressively drawn attention since it can improve separation of biodiesel and glycerol, leading to the reduction of production costs. Also, it can take the reaction in favor of more productive products by selective separation of biodiesel from reaction media (Shahid and Jamal, 2011; Sdrula, 2010).

Although many types of oil feedstock could be converted into biodiesel through the transesterification reaction, lack of edible oil for human consumption in developing countries does not approve its usage for biodiesel production. Moreover, about 70% to 88% of the total biodiesel production cost ascends from the price of raw material (Babazadeh et al., 2017). Thus, lowering the production costs is possible through changing the feedstock from edible oils to non-edible oils. Non-edible oils, such as Jatropha, bitter almond, Mahua, and many others, animal fats, and waste cooking oils have been used as feedstock that could lower the expenses. As a result, non-edible oils, like waste cooking oils (WCO) and refinery waste oil, have been increasingly recommended for biodiesel production. Today, growing food consumption has improved the production of a large amount of such oils/fats globally. For example, residual olive oil recovered from olive cake is 0.11- 0.22 billion liters a year (Abdullah et al., 2019; Bhuiya et al., 2014).

Figure 1 shows the block flow diagram of biodiesel production from vegetable oils and waste cooking oils containing high content of Free Fatty Acid (FFA). Transesterification is the main part with a pretreatment step (esterification reaction) for converting feedstock with a high content of FFA to FAME through esterification reaction in the presence of an acid catalyst like sulfuric acid.

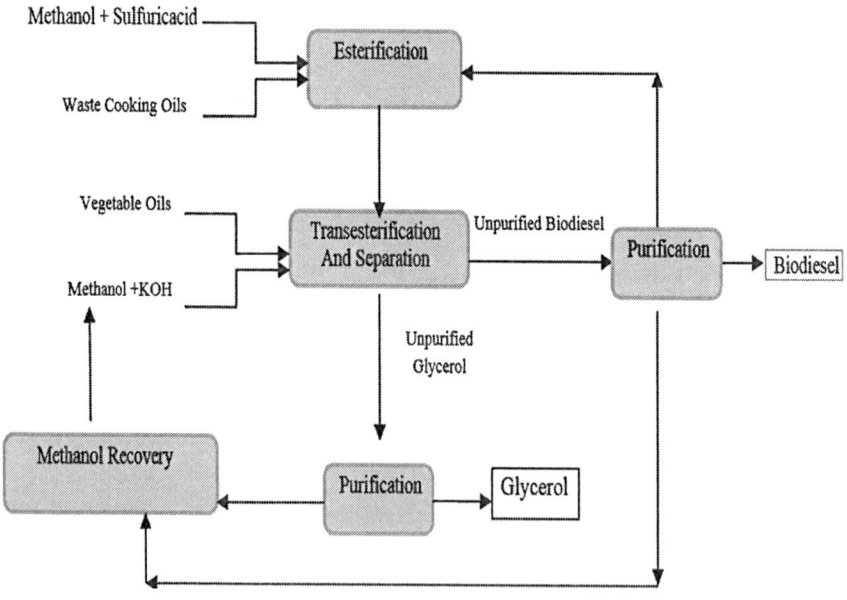

Figure 1. Block Flow Diagram of biodiesel production process from vegetable oils and waste cooking oils with high amount of FFA.

In general, the reaction rate will be accelerated in the presence of a catalyst. If this catalyst is in the same liquid phase to that of reactants phase, it is called homogeneous catalytic reaction; otherwise, if the catalyst is in a different phase, i.e., solid or immiscible, to that of reactants, we have heterogeneous catalytic reaction (Knothe, Krahl, and Van Gerpen, 2015). In this regard, all types of catalyst used for transesterification of vegetable oils can be divided into three categories of homogeneous, heterogeneous, and enzyme. Enzyme catalysts can also be divided into either heterogeneous or homogeneous, depending on mobility characteristics. Alkaline liquid catalysts, like sodium or potassium hydroxide, are of homogeneous base catalysts and acidic liquid catalysts, like sulfuric acid, are of homogeneous acid catalysts. Furthermore, solid base or acid catalysts, like titanium silicate, compounds of alkaline-earth metal, anion exchange resins, and sulfated metal oxides, are classified as heterogeneous catalysts. Lips or Novozymes 435 are among enzyme catalysts (Buyukkaya, 2010; Pinto et al., 2005).

Homogeneous catalyst systems used for biodiesel production through transesterification reaction have problems, such as base catalyst neutralization step, production of a large amount of wastewater during the separation steps, and emulsification possibility (causing emulsified biodiesel with glycerin and making the separation more difficult, especially if the alcohol is ethanol) (Buyukkaya, 2010; Pinto et al., 2005). This is while several commercial plants producing biodiesel from vegetable oils by transesterification use homogeneous base catalyst (Semwal et al., 2011). Homogeneous catalysts can be replaced by heterogeneous catalysts (solid catalysts) because of their advantages, such as easy separation, less equipment corrosion, less environmental problems (Yan et al., 2011), elimination of neutralization step, higher activity and selectivity, and longer lifetime (Ilgen, 2011). Meanwhile, catalyst recyclability and production of much less wastewater during the process make them to be considered as Green Technology (Chouhan and Sarma, 2011). It is noteworthy that solid catalysts should be introduced to reduce the overall cost of the biodiesel production process by less and easier separation steps (Chouhan and Sarma, 2011).

So, the use of heterogeneous catalysts emerges as a preferred route (Chouhan and Sarma, 2011). Hence, various types of catalysts, such as alkali-earth metal compounds, sulfated metal oxides, sulfated zirconia, mixed metal oxides, supported heteropolyacids, heterogenized metal cyanide, vanadyl phosphate, zirconia, tin compounds supported in ion-exchange resins, alkyl guanidine heterogenized on organic polymer, zeolite, alumina loaded with alkali metal salts, and calcium oxide loaded with lithium-ion have been reported so far. Recent studies investigated functionalized mesoporous silica, such as tin oxide modified mesoporous SBA-15, titanium-grafted mesoporous silica, and magnesium supported MCM-41. However, high temperature and pressure are needed for their complete conversion (>98%). Generally, heterogeneous catalysts require longer reaction time than homogeneous catalysts due to the formation of three phases of alcohol, oil, and catalyst, causing limitation problem that can be overcome using precise amount of co-solvent as a solution (Chouhan and Sarma, 2011).

Several factors, such as pore size, pore volume, specific surface area, and active site concentration affect catalyst efficiency (Trombettoni et al., 2018). It is also believed that the properties of an ideal catalyst are: 1) large enough pore size to minimize diffusion problem, 2) high concentration of active site, 3) high catalytic stability against poisoning and leaching, and 4) possibility to tune hydrophobicity of surface to enhance the selective adsorption of substrate and repulsion of greatly polar compounds, which can cause deactivation. The pore size of the catalyst is important due to the fact that most of organic molecules are large. Therefore, materials with controlled pore size, such as silica-alumina, has attracted great attention (Chouhan and Sarma, 2011).

Some of the characteristics of biodiesel, like quick biodegradability, low emission profiles, and nontoxicity, make it a perfect fuel for environmentally sensitive applications as underground mining, urban buses, and marine areas. Biodiesel is known as the alkyl ester of fatty acids derived from renewable resources. Hydrogenated vegetable oils, animal fats, and some tropical oils, such as coconut and palm oil, are not preferred as feedstock for conversion to biodiesel. Because they contain 35–45%

saturated fatty acids and tend to gel even at relatively high temperatures. Oilseeds, such as rapeseed or soybean, are recommended for conversion to biodiesel.

The handiness of biodiesel feedstock is diverse in each region. Climate and geographical conditions and agronomic reasons are the main factors leading to the productivity of seeds like soybean or rapeseed. Europe, Southeast Asia, and North America are great biodiesel users due to their production capacity of vegetable oils.

BIODIESEL BIODEGRADABILITY

Biodiesel can be termed clean fuel due to the lack of carcinogens and less Sulphur contents. Biodiesel is an attractive choice since it is concerned with environmental protection and ecosystem health. Biodiesel is a locally available, non-polluting, sustainable, accessible, and reliable fuel obtained from renewable sources. It generally contributes to the reduction of CO and HC emissions compared to those of fossil-based engine fuels. Biodiesel fuels have also the benefit of being renewable fuels (Knothe, Krahl, and Van Gerpen, 2015).

Biodegradation is degradation caused by biological activity, particularly with enzyme action leading to significant changes in the material's chemical structure. Biodegradability of biodiesel has been proposed as a solution for the waste problem to see whether they can be a local environmental hazard in accidental discharges or not. There are many methods, such as carbon dioxide (CO_2) evolution, gas chromatography (GC) analysis, and biochemical oxygen demand (BOD) measurement, for biodegradation (Estill, 2005).

Lots of experiments have been done on biodegradability of biodiesels. It is concluded that biodiesels are "readily biodegradable" compounds according to Environmental Protection Agency (EPA) standard (EPA, 1982). Biodiesel is a natural product consisting of pure fatty acids. During the degradation procedure, the enzymes are responsible for the breakdown of biodiesels (Sarin, 2012). All fatty acids are even hydrocarbon chains in ester that are formed with two linked oxygen atoms that make them very biologically active. The natural enzymes recognize oxygen atoms and attack them instantly and the degradation process occurs. During this process, fatty acids are degraded to acetic acid and a fatty acid with two fewer carbons. However, diesel has a more complicated chemical composition. It consists of a large amount of alkane and alkene without any attached oxygen that are not biologically active, such that the natural enzymes may not recognize them (Fang, 2012). Diesel also contains aliphatic cyclic hydrocarbons, polycyclic aromatic hydrocarbons (PAHs), and alkylbenzenes that are toxic to microorganisms. Moreover, Benzene needs more energy for microorganisms to open the ring since it is highly stable (Sreenath and Pai, 2018).

Biodegradability of various biodiesel fuels in aquatic and soil environments was researched by the University of Idaho in the mid-1990s. The study was done by the CO_2 evolution method, gas chromatography (GC) analysis, and seed germination using neat oils and biodiesel from a variety of feedstock, including soy, canola, rapeseed, and others. Both methyl and ethyl esters were included. Phillips 2-D was used as reference petroleum diesel. Blends of biodiesel/petrodiesel at diverse volumetric ratios were also included. CO_2 evolution results for all biodiesel fuels reached 84% in the aqueous system, and the average substance disappearance was 88% in the soil environment. The results showed that the presence of biodiesel in the blend increases the extent of biodegradation of petrodiesel up to 100%. The seed germination test showed that a biodiesel fuel spill-contaminated soil can be restored by biodegradation in 4–6 weeks to such a degree that it can support plant germination. However, biodiesel-contaminated soil did have an effect on plant growth for the first 3 weeks due to the rapid growth of microorganisms during fuel degradation (Sharp, 1996).

Biodiesel can also promote and speed up biodegradation of diesel. Greater amount of biodiesel in a biodiesel/diesel mixture results in faster rate of degradation. This is because microorganisms metabolize both biodiesel and diesel at the same time and almost the same rates [42].

Biodegradation of polluted soil with diesel, petroleum, and biodiesel was investigated by Cruz, showing that native microorganisms can degrade soil pollutants. Their results showed that the soil polluted with petroleum and diesel was little active. It is because of the complex composition of petroleum and diesel, which requires a microbial succession to biodegrade hydrocarbons. It was observed that biodiesel was completely consumed in the first 120 days of soil incubation. The test of oil residual revealed that there was no biodiesel in the incubated soil during 240 days, but there was petroleum in soil after 240 days in contact with soil microorganisms. Due to the presence of polycyclic aromatic hydrocarbons (PAH) with several aromatic rings, biodegradation of petroleum by microorganisms is difficult, which was described by Atlas and Nwaoguetal (1995).

Zhang et al., (1998) showed that vegetable oils are slightly less degraded than their modified methyl ester. Many vegetable oils comprise the polyunsaturated fatty acid chains that are methylene interrupted instead of conjugated. The double bond of unsaturated fatty acids limits rotation of the hydrogen atoms linked to them. Therefore, an unsaturated fatty acid with a double bond can be in two forms. In the cis form, the two hydrogens are on the same "side" and in the transform, the hydrogen atoms are on the opposite sides.

Biodegradability of different biodiesel was compared with the conventional reference diesel fuel (2-D) by Demirbas (2008). The obtained results are in good agreement with earlier observations and demonstrate that all the assessed biodiesels are biodegradable, with curve-like shapes of 80.4 to 91.2% biodegradation after 30 days, while the reference diesel fuel sample touched merely 24.5% biodegradation. High degradation rates were presented over the first 10 days (Zhang et al., 1998).

The variance between the biodegradability rates of biodiesel samples can relate to the diverse structures of fatty acids in the vegetable oils. The linoleic and linoleic acid contents in vegetable oils are significantly different, even if the present chemical compounds are the same. Biodegradation had a much greater effect on the polyunsaturated fatty acids, such as linoleic and linoleic acids, than saturated and mono-saturated fatty acids (Zhang et al., 1998).

TOXICITY

Biodiesel does not comprise any crude oil residues or toxic metals; therefore, it does not contribute to any toxicity as petroleum diesel (Venu, Subramani, and Raju, 2019).

The toxicity of some biodiesels was investigated and compared with Phillips 2-D low-sulfur diesel fuel by the University of Idaho in the mid-1990s. The case examples of the study included rapeseed methyl ester (RME), rapeseed ethyl ester (REE), and their mixtures with each other and with Phillips 2-D low-sulfur diesel fuel in different rates. The results of the acute oral and acute dermal toxicity tests showed that biodiesel is avoidable to be ingested and it is not allowable to be contacted with skin because of some adverse effects. The animals treated with diesel had more injurious clinical observations (Knothe, Krahl, and Van Gerpen, 2015).

The 100% RME fuel was the slightest severe toxic fuel in the acute oral toxicity study, and the 100% REE was the minimum severe toxic fuel in the acute dermal toxicity study (Knothe, Krahl, and Van Gerpen, 2015).

Biodiesel is not as toxic as D. magna and NaCl. Compared to the reference toxicant (NaCl), diesel fuel was around 2.6 times more toxic, RME was 6.2 times less toxic, REE was 26 times more toxic, and SME was 89 times more toxic. Compared with the number two diesel fuel, RME, REE, and SME were16, 69, and 237 times less toxic.

In both the static and flow-through tests, the rapeseed-based fuels, REE and RME displayed the highest EC50 values, signifying that they are less toxic than other test substances. The EC50 values of other vegetable oil-based fuels, SME, were lower than rapeseed fuels (significance untested). The results of the static test for SME failed the $\chi 2$ test for heterogeneity using a Probit model. The REE and 2-D mixture results were as expected, i.e., 20/80 mixture had higher EC50 values than the 50/50 mixture for both the static and flow-through analyses. This approves the results of other tests that indicated a higher toxicity (lower EC50) for a higher percentage of 2-D in the mixture and a lower toxicity (higher EC50) with an increasing percentage (Knothe, Krahl, and Van Gerpen, 2015).

ENVIRONMENTAL IMPACTS

Biodiesels lower greenhouse gas (GHG) emissions. Moreover, they provide high oxygen content and aromatics, low sulfur, and water content. Many studies have been done on the reduction of the amount of exhaust gases, achieving remarkable results in the last decade. Some of these studies are mentioned here.

Nabi et al., (2009) used cottonseed oil and indicated that 30% of biodiesel mixtures diminished CO emissions by around 24%. The NOx emission was increased around 10% during the use of the biodiesel. PM emission was decreased around 20% during the use of 20% biodiesel. Also, 10% biodiesel could decreased around 14% of smoke emission.

Karanja (Pongamia Pinnata) was assessed in the study of Nabi et al., (2009), proving that the NOx emission was increased, as the previous study, during the use of biodiesel, but other factors decreased. When they just used biodiesel, the CO and smoke emissions decreased more than biodiesel mixtures, but the NOx emission enhanced (50%, 43%, and 15%).

Jatropha, Karanja, and polanga were used by Sahoo et al., (2009), who showed that the growth of biodiesel in diesel reduced smoke, hydrocarbon, and particulate matter noticeably. However, they found that CO, NOx, and HC promoted slightly.

Karabektas et al., (2009) applied turbocharger and observed that CO emission declined during the use of biodiesel, but as the previous study, NOx emission increased. These results were in line with Charles et al., who indicated that the amount of NOx increased with the increased rates of biodiesel in the mixtures (Mueller, Boehman, and Martin, 2009).

Bolszo et al., (2009) operated the 30 kW gas turbine engine on biodiesel. The results were focused on NOx emissions, and showed that increased biodiesel increased NOx emission.

Macor et al., (2009) investigated boiler fueled with biodiesel and heating oil. The CO and PM were diminished during the use of biodiesel compared to heating oil. They also reported that PAHs contained in the PM were around 13 times less toxic compared to oil case. They also found that the amount of formaldehyde was 2 times lower for the biodiesel and VOCs in both cases.

Di et al., (2009) used the ultra-low sulfur diesel in their study and proved that the amount of NOx and NO2 emissions increased during the decrease of HC and CO emissions. At high loads, both smoke and PM concentrations were significantly reduced. The results showed that the combination of biodiesel and ultra-low sulfur diesel depicted similar outcomes to those in the literature applying higher sulfur diesel fuels and biodiesel from other sources.

Li et al., (2009) indicated that the amount of CO diminished by 20% with the use of Eruca Sativa Gars vegetable oil. They showed that the amount of CO_2 and NOx also increased. HC reached 0% and smoke opacity decreased with the use of neat biodiesel.

Lir et al., (2009) applied rice bran oil, resulting in a remarkable decrease in CO, HC, and PM, but a slight growth in NOx.

Krah et al., (2009) employed rapeseed biodiesel and showed that the decrease of HC, CO, PM except NOx emissions increased up to 15% with the use of biodiesel.

Soybean biodiesel was surveyed by Yoon et al., (2009), who compared biodiesel and diesel emissions and found that the biodiesel fuel emitted higher NOx and an extraordinarily low level of soot emissions.

Buyukkaya (2010) applied rapeseed biodiesel in their study. They found that smoke opacity decreased by around 60% and CO was diminished by about 32%. As the previous studies, the amount of NOx emission was enhanced during the growth of the engine speed with the use of biodiesel, which was higher than diesel fuel.

In the study of Zhu et al., (2010), reduction of NOx and particulate with the use of biodiesel blends was investigated. They indicated that biodiesel could decrease NOx and particulate compared to Euro V diesel fuel and they found that biodiesel-methanol was more effective than biodiesel–ethanol. Also, they indicated that the increase of alcohol in the blends led to greater reduction of NOx and particulate. The increase of alcohol in blends caused the CO and HC increase.

Barabás et al., (2010) showed that CO emission diminished by around 59% and CO_2 enhanced with the use of biodiesel. The amount of NOx emission grew slightly and the amount of smoke and HC diminished at high loads.

Aydin et al., (2010) applied ethanol as the additive to investigate the use of greater percentages of biodiesel in an unmodified diesel engine. Diesel and biodiesel-ethanol enhanced the amount of NOx emission slightly, but the amount of CO_2 diminished by around 67% with the use of biodiesel-ethanol. They also mentioned that the amount of CO and SO_2 emissions decreased by increasing the biodiesel in blends.

In the study of Kim et al., (2010), the CRDI diesel engine was equipped with warm-up catalytic converters or the catalyzed particulate filter. They indicated that biodiesel diminished HC, CO and increased NOx emission due to the oxygen content. The smoke emission decreased by about 50%. They also found that the total number of particles decreased; however, biodiesel–diesel blends emitted more particle, which was under 50 nm in comparison to neat biofuel.

Ryu et al., (2010) surveyed the effect of antioxidants on exhaust emissions of the diesel engine. They indicated that the antioxidants had an insignificant effect on the exhaust emissions.

The apricot seed kernel oil was used as biodiesel in another study. They found that the inferior concentration of the apricot biodiesel showed a higher improvement in the exhaust emissions. Consequently, minor percentages of apricot biodiesel can be applied as additive (Gumus and Kasifoglu 2010).

Tsai et al., (2010) showed that the emission concentrations of OC, EC, and TC were at the bottom with the use of B10 or B20. Both B10 and B20 emitted total-PAHs by around 38% and 28% with the use of soy-biodiesel, which could decrease the PM, carbon.

The results of the study by Bakeas et al., (2011) was different with the use of four different low biodiesel concentrations. They concluded that the amount of NOx emission of diesel fuel was the same as biodiesel fuel. The presence of oxygen in biodiesel blends caused reduction in PM. Also, CO and HC increased over the cold-start.

Behçet (2011) investigated waste anchovy fish biodiesel and showed that the amount of HC, CO_2, and CO reduced by 33.42%, 4.576%, and 21.3% during the process. The amount of NOx and O2 increased by about 29.37% and 9.63% with the use of biodiesel. The smoke opacity was also reduced by about 16% in comparison to diesel fuel.

Jatropha biodiesel was applied by Ganapathy et al., (2011), who investigated the effect of injection timing on smoke, CO, and HC. The results showed that it decrease smoke, CO, and HC by 1.5%, 2.5%, and 1.2%, respectively.

The waste cooking oil methyl ester was used in a study conducted by Muralidharan et al., (2011). The results of this study showed that biodiesel met the standards of the diesel fuel parameters. The blends decreased CO_2, CO, and HC at the expense of nitrogen oxides emissions.

Kannan et al., (2011) surveyed the use of ferric chloride, as fuel borne catalyst, for waste cooking palm oil. The fuel borne catalyst added to biodiesel resulted in lower NO and increased CO_2 emissions compared to diesel fuel. Also, CO, HC, and smoke diminished by around 52.6%, 26.6%, and 6.9%, respectively.

Hulwar et al., (2011) applied ethanol in diesel–ethanol blends with biodiesel and showed that at high loads, the smoke opacity decreased highly. Also, NO variation relied on operating situation, while the CO emissions increased considerably.

Qi et al., (2011) worked on ethanol and diethyl ether as additives to biodiesel. They showed that ethanol and diethyl ether resulted in smoke reduction. They also showed that NOx and HC were increased, but CO was decreased.

The neat karanji oil was used in the study of Anand et al., (2011). They found that HC and CO emissions were higher with the use of methanol blend compared to neat biodiesel at low load situation. This is while at high load situation, HC emissions were similar for the two fuels, and carbon monoxide emissions considerably diminished with the use of methanol blends. An outstanding reduction in NO and smoke emissions was observed with the use of biodiesel-methanol blend.

Vedaraman et al., (2011) used palm oil in their study and showed that B20 decreased the amount of HC and CO by around 28 and 30% in comparison to diesel. The NOx emission in this study was the same for biodiesel and diesel fuels, which was reduced by 2% by adding distilled water.

Jatropha biodiesel was used in a study by Tan et al., (2012), who showed that smoke and HC emissions experienced a continuous decreasing trend with growing biodiesel blend ratios. At low engine loads, CO emission of biodiesel was increased. Also, carbonyl compound emissions was increased when the whole engine was operating. Biodiesel decreased aromatic hydrocarbon emissions of diesel engines.

Another study on Jatropha biodiesel by Chauhan et al., (2012) indicated that Jatropha biodiesel decreased smoke, CO_2, CO, and HC emissions. This is while the amount of NOx emissions was increased in comparison to diesel fuel.

Yang et al., (2012) surveyed the emission of biodiesel at partial load situation. The results of this study showed that CO emission at low engine loads increased during the increase of the biodiesel ratio, which was vice versa at high load conditions.

Gumus et al., (2012) studied the impact of fuel injection pressure on exhaust emissions with the use of biodiesel. They showed that the amount of CO_2, NOx, and CO emissions increased. On the contrary, HC, CO, and smoke emissions experienced a decreasing trend.

Wahlen et al., (2012) applied yeast (Cryptococcus curvatus), bacteria (Rhodococcus opacus), and microalgae (Chaetoceros gracilis) in their study and showed that HC and CO emissions were significantly lower with the use of biofuels compared to diesel fuel. Also, biofuels emitted lower NOx in comparison to diesel fuel.

Yang et al., (2013) applied waste cooking oils and showed that HC and NOx were reduced at high loads with the use of biodiesel. On the contrary, opposite results were obtained at low engine loads. Their results proved that partial and high engine speeds affect gas emissions.

Pinzi et al., (2013) investigated the effect of fatty acid composition on emission and showed that the impact of bulk modulus on NOx was higher compared to shorter fatty acids. Volatile organic fraction, HC, and CO were higher during the enhancement of FAME chain length. Soot formation was mostly influenced by viscosity and oxygen content of FAME.

Liaquat et al., (2013) used coconut biodiesel and showed that less CO and HC, and more NOx and CO_2 emissions were observed in biodiesel blended fuels in comparison to diesel fuels.

Özener et al., (2014) employed soybean biodiesel to investigate CO and HC emissions that were reduced by 28-46%. NOx and CO_2 emissions were enhanced by about 1.46–5.03 due to the oxygen content.

Yilmaz et al., (2014) studied the effect of biodiesel–butanol fuel blends on gas emissions. The results showed that butanol blended fuels can reduce exhaust gas temperatures and NOx emission, while increasing HC and CO emissions. Butanol blended fuels generated higher NOx and lower CO emissions in comparison to diesel fuel with low butanol concentrations (5% and 10%), while there was no significant change in terms of HC emissions. A biodiesel blend with the maximum butanol concentration (20%) resulted in higher HC and CO emissions and less NOx emission in comparison to diesel fuels.

Fattah et al., (2014) used palm biodiesel to investigate the impact of antioxidants on emissions. Both antioxidants reduced NOx emission by about 9.8–12.6% in comparison to B20, but increased CO and HC by around 8.6–12.3% and 9.1–12.0%, respectively.

The waste cooking oil used in the study of Can (2014) caused an increase of about 8.7% in NOx emissions. Smoke and HC were also decreased with the use of biodiesel. The amount of CO emission was not significantly changed at low and medium engine loads, but it decreased at high load engine. Finally, CO_2 was slightly increased at all engine loads.

In the study of Mohsin et al., (2014), diesel dual fuel was applied for reducing exhaust emissions. Biodiesel significantly increased carbon monoxide by about 15–32% and NO by about 6.67–7.03%; however, it reduced HC and CO_2 by about 5.76–6.25% and 0.47–0.58%, respectively.

Fattah et al., (2014) applied Calophyllum inophyllum biodiesel blends accompanied by oxidation inhibitors. Results of this study showed that

CIB20 increased NOx, but decreased HC and CO emissions. Antioxidants decreased NOx emission by 1.6–3.6%, while increased both HC and CO emissions in comparison to CIB20. However, it was under the diesel emission level.

In a study by Li et al., (2015), pentanol was added to biodiesel, resulting in the decrease of soot emission. At middle loads, blends could decrease the amount of NOx emission compared to diesel fuel. At low engine loads, CO and HC increased with the use of pentanol blends.

Mirzajanzadeh et al., (2015) used Nano-catalyst in their study to investigate the effect of hybrid nanocatalyst comprising cerium oxide on amide-functionalized multiwall carbon nanotubes. Their results showed that HC, NOx, soot, and CO emissions were decreased by about 71.4%, 18.9%, 26.3%, and 38.8%, respectively.

Canola oil–hazelnut soapstock biodiesel mixture was employed in a study by Öztürk et al., (2015), who showed that B5 oxygen content surged the combustion yielding to increase NOx emission and decrease THC, smoke, and CO emissions. While, B10 fuel decreased combustion due to higher viscosity, density, and surface tension. Consequently, smoke, THC, and CO emissions were increased, but NOx emission was reduced. Also, CO_2 emissions with the use of both biodiesel blends were similar to those of the diesel fuel.

Omidvarborna et al., (2015) investigated NOx emission from low-temperature combustion of biodiesel prepared with different feedstock and blends. They found that unsaturated fatty acids caused the rise of NOx emission compared to biodiesel with high saturated fatty acids.

Pentanol and Calophyllum inophyllum biodiesel blends were employed in a study by Imdadul et al., (2016), who showed that pentanol resulted in lower NO, CO, and HC emissions in comparison to Calophyllum inophyllum. At high engine speed, pentanol resulted in greater reduction of smoke and CO_2 in comparison to Calophyllum inophyllum.

The waste cooking oil was used in the study of Cheikh et al., (2016). They showed that UHC increased with engine speed increase, indicating the great effect of load. Also, PM and CO emissions decreased, depending on the load, and were less sensitive to engine speeds.

The blend of palm and Jatropha curcas was applied in the study of Nalgundwar et al., (2016), showing that CO emission was decreased by around 14.5%, 17.7%, and 7.1% with the use of B5, B10, and B20 in comparison to diesel fuel. However, B5 and B10 increased NOx emission by around 5.3% and 9.2%, respectively.

Zheng et al., (2016) used biodiesel/n-butanol, biodiesel/ethanol and biodiesel/2,5-dimethylfuran and showed that biodiesel/ethanol decreased NOx emission compared to diesel fuel. At low loads, all samples showed higher CO and HC emissions compared to diesel fuel. Biodiesel/n-butanol and biodiesel/2,5-dimethylfuran could decrease the amount of soot emission by about 79% and 99.4%, respectively.

Soybean biodiesel fuel was examined in the study of Car et al., (2016) at high engine loads. The results of this study showed that smoke and NOx emissions were simultaneously increased up to 15% and 55%. Also, THC emissions declined at medium and low engine loads. Although CO emission changes were insignificant, it was increased at high engine load. Nevertheless, CO_2 emissions were higher at all engine loads.

Prabu and Anand (2016) added alumina and cerium oxide nanoparticles to biodiesel. They found that all NO, CO, HC and smoke emissions were reduced by 13%, 60%, 33%, and 32%, respectively.

In the study of Gnanasekaran et al., (2016), fish oil reduced HC, NOx, and CO emissions compared to diesel fuel. This is while biodiesel increased smoke emissions. Retardation of injection timing reduced the of emission parameters, such as NOx, HC, and CO.

Man et al., (2016) used waste cooking oil and showed that the growth of biodiesel in the blended fuel reduced CO, HC, and particulate mass concentrations, while increased NOx. For unregulated emissions, acetaldehyde and formaldehyde emissions were increased with growing biodiesel content. For aromatics emissions, biodiesel addition contributed to an increase in benzene emission, but to a decrease in xylene and toluene emissions. The results showed that all emissions are influenced by engine operating modes, particularly the engine load.

Radhakrishnan et al., (2017) worked on pentanol and palm biodiesel to investigate greenhouse gas emissions. They found that blends could decrease HC, NOx, smoke, and CO by about 3.8%, 6.6%, 2.7%, and 9.3%, respectively.

Devarajan et al., (2017) examined the cashew nutshell biodiesel and the results showed that CO, HC, Smoke, NOx emissions were decreased by about10.1%, 2.6%, 2.1%, and 5.1% by adding about 10% of pentanol to biodiesel. Besides, they found that there were no modifications in the engines with the use of these blends.

TiO_2 nanoparticle in the powder form was added to mahua biodiesel in the study of Pandian et al., (2017). They showed that emissions can be reduced by adding nanoparticle to biodiesel. Also, CO, HC, Smoke, and NOx emissions were decreased by about 9.3, 5.8, 2.7, and 6.6%, respectively.

Nabi et al., (2017) worked on the effect of biodiesel, derived from waste cooking oil, on emissions. Their results showed significant drops in both particulate matter and number emissions. Also, CO and HC emissions were reduced with the use of biodiesel, while NOx emissions experienced a minor increase.

Gharehghani et al., (2017) used waste fish oil biodiesel to investigate its impact on gas emissions. The results showed that CO emission concentration was declined with a mild slope (5.2–27%), whereas outstanding reduction was observed in UHC (11.6–70%). Oxygen content of biodiesel contributed to higher efficiency of combustion, resulting in about 7.2% growth in CO_2. This is while NOx emission was enhanced (1.9–12.8%) with the use of biodiesel and blends.

Alptekin (2017) examined diesel fuel, canola-safflower biodiesel, and biodiesel additives, such as solketal and ethanol. He found that biodiesel emitted greater CO_2 and NOx, but lower THC and CO compared to diesel fuel. Moreover, ethanol and solketal fuel blends caused an increase in NOx emissions and showed lower CO, CO_2, and THC emissions on average compared with biodiesel. Solketal-biodiesel and ethanol-biodiesel had almost the same results.

Tan et al., (2017) investigated the effect of diesel-biodiesel-bioethanol emulsion fuels on emissions. Their results showed a decrease of CO and CO_2 in comparison to diesel fuel. This mixture decreased NOx emission at medium engine speeds by about 30%. At low and high engine speeds, NOx emission was increased.

Polyoxymethylene Dimethyl Ethers was added to the biodiesel in a study by Li et al., (2017), who showed that this resulted in lowermost soot emissions and could halt the trade-off between soot and NOx emissions. Remarkably, the emissions of soot and NOx reached 0.0009 g/kW h and 0.05 g/kW h, respectively.

Pandian et al., (2017) used TiO2 as additive to mustard oil biodiesel. Experimental data showed that nanoparticles (TiO_2) had a positive effect on biodiesel emission as it worked as an oxidation buffer. This caused a decrease in smoke, HC, and CO emissions in comparison to pure biodiesel. NOx emissions were also declined with this catalytic, which decreased the peak combustion temperature. Thus, TiO_2 nano-fluid had an obvious effect on decreasing neat biodiesel emissions.

Joy et al., (2018) employed octanol and biodiesel blends and examined ther impact on gas emissions. They found that at all loads, CO emission was lower with the use of biodiesel and biodiesel–octanol in comparison to diesel fuel. Also, the amount of CO was lower with the use of biodiesel–octanol due to higher oxygen content. HC emission was the same as CO emission. In addition, NOx emission was lessened with the use of diesel, while biodiesel–octanol had a lower level of NOx emission in comparison to neat biodiesel. Finally, smoke emission was the same as CO and HC and was lower with the use of biodiesel–octanol.

Mahalingam et al., (2018) added octanol to neat mahua oil biodiesel. They showed that CO emissions declined with increased octanol ratios in the blends. A 6.8% and 7.4% decrease in total CO emissions were was observed with the use of B20 and B10 at all loads in comparison to neat mahua oil biodiesel. HC had the same condition as CO and decreased by about 5.1% and 5.7%. Also, NOx emission had a decline of about 4.8% and 5.4% as CO. Smoke emissions declined with increased octanol ratios in the blends. Finally, 2.1% and 2.9% decrease in total CO emissions was observed with the use of B20 and B10 at all loads in comparison to neat mahua oil biodiesel.

Dahs et al., (2018) worked on ethyne gas in biodiesel-fueled diesel and showed that the maximum decrease in smoke emission (24.8%), CO (17.24%), and HC (25.1%) was observed in biodiesel–ethyne (at 3 L/min) in comparison to neat biodiesel. This is while, NOx emission was 15.8% higher for ethyne–biodiesel fueling (at 3 L/min) due to increased combustion gas temperature in comparison to neat biodiesel.

Mahalingam et al., (2018) investigated the effect of pentanol-biodiesel blend derived from mahua oil on emissions. Smoke and NOx emissions were decreased by about 5.1–6.4% and 3.3–3.9 with biodiesel and pentanol blends in comparison to neat biodiesel. Also, HC and CO emissions decreased by about 2.1–3.6% and 3.1–4.2% with the addition of pentanol to biodiesel.

Rathinam et al., (2018) added water particles of various proportions to orange peel oil biodiesel to examine its effect on gas emission. In the presence of water- biodiesel, HC and CO emissions were less in comparison to neat biodiesel. The highest decrease was about 10.1 for CO and 8.8% for HC, which was due to higher evaporation tendency. NOx and smoke emissions were around 12.4 and 18.4% for water-biodiesel in comparison to neat biodiesel, which was due to the reduction of peak temperature inside the cylinder.

Devarajan et al., (2018) added N-Octanol to Mustard oil biodiesel to investigate its effect on gas emission. Due to the oxygen content, n-octanol affected gas emissions. N-octanol was a better oxidizing catalyst and greatly decreased CO and HC emissions. A significant decrease in NOx emission was observed in case of fueling with emulsion method.

Radhakrishnan et al., (2018) examined the effect of adding alumina nanoparticles to biodiesel (cashew nut shell oil). Experimental data on unmodified engine showed that the emission parameters, such as CO, HC, smoke, and NOx, decreased by 5.3%, 7.4%, 16.1% and 10.23% for BD100%, and 8.8%, 10.1%, 18.4% and 12.4% for B100A, respectively.

Devarajan et al., (2018) studied neat neem oil methyl ester in the presence of silver oxide nano-particles. Emissions of CO, HC, smoke, and

NOx were decreased by 12.22%, 10.89%, 6.61%, and 4.24% for BD100+ Ag2O and 16.47%, 14.21%, 8.34%, and 6.66% for BD100, respectively. Also, in another study by Devarajan et al., (2018), who added Ag2O to palm oil, HC, CO, smoke, and NOx emissions were increased.

Zhang et al., (2018) investigated the effect of fatty acid methyl esters on four biodiesels. They found that the ignition delay time and kinematic viscosity play a key role in biodiesel emissions. The chemical ignition delay time can be reduced by higher saturation level and also the kinematic viscosity can be increased by greater saturation contents, such as C18:0 and C16:0 together with the C18:1, which was the single double bond methyl ester. The increased kinematic viscosity can contribute to poor evaporation procedure and poor fuel-air mixing. Minor kinematic viscosity, such as C18:3 and C18:2, was helpful for better combustion and fuel-air mixing, but resulting in greater NOx emission. Therefore, the connection between combustion and emission characteristics and also the proportion of biodiesel was not straightforward and easy to determine. Accordingly, the balance of five components of biodiesel fuel was of utmost importance. Compared to pure diesel, biodiesel oxygen content increased in-cylinder combustion. It was also useful to decrease CO and HC emissions and increase NOx emissions. Though, it was not clear at low load.

Uyumaz (2018) examined mustard oil biodiesel and found that smoke and CO emissions were remarkably reduced with the use of biodiesel, while increasing NOx emission. NOx emission was 582 ppm with D100, which was increased by about 22.1%. Results showed that low levels of mustard oil biodiesel appears to be optimal when the engine operated at partial engine loads.

Jatropha biodiesel and decanol additive were investigated in a study by Balan et al., (2019). Results of this study showed that biodiesel can increase NOx emission by about 10.65%. However, in the presence of 10% decanol and biodiesel, NOx emission was decreased by about 6.8% and 5.3% in comparison to biodiesel and diesel fuel. Increasing the amount of decanol by about 20% resulted in 7.5% and 4.7% decrease of NOx compared with biodiesel and diesel fuel. The reason for this is the lower calorific value. Also, HC emissions decreased by about 10.65% compared with diesel fuel. But, in the presence of 10% decanol and biodiesel, HC emission was decreased by about 5.3% and 6.8% in comparison to biodiesel and diesel fuel. Increasing the amount of decanol by about 20% resulted in 17.1% and 19.6% decrease of HC compared with biodiesel and diesel fuel. The CO emission decreased by about 8.8% with the use of biodiesel in comparison to diesel fuel. But, in the presence of 10% decanol and biodiesel, CO emission was decreased by around 18.1% and 16.5% in comparison to biodiesel and diesel fuel. Increasing the amount of decanol by about 20% resulted in 17.1% and 19.6% decrease of CO compared with biodiesel and diesel fuel. The reason for emission decrease was the lavish availability of oxygen in decanol-biodiesel blends.

In the study of Yithya et al., (2019), who used TiO_2 and canola biodiesel, it was found that the exhaust emission of CO, HC, and NOx was reduced. The canola biodiesel blends resulted in inferior NOx and lower smoke emissions.

The study of Joy et al., (2019) evaluated the impact of di-Methyl-Ether, as cetane reformer on neat cashew nut shell biodiesel, on gas emission. The results showed that in the presence of 20% of Di-Methyl-Ether, CO, HC, NOx, and smoke were decreased by around 3.4%, 4.2%, 8.8%, and 8.4%, respectively.

Siva et al., (2019) worked on decreasing NOx and smoke emissions under the effect of water addition to waste orange peel oil biodiesel. They found that HC and CO were considerably decreased with the use of additive in the biodiesel. Also, HC decreased by around 12.2% for 4% water ratios and 16.3% for 8% additive. The amount of CO emission was decreased by 17% and 21.8% with the use of 4% and 8% water ratios, respectively.

Devarajan (2019) used di-methyl-carbonate as cetane improver on neat almond biodiesel. Findings suggested that 20% of di-methyl-carbonate decreased HC by 5.2%, CO by 7.4%, smoke by 3.6%, and NOx by 4.7% with the use of BD100.

Neem biodiesel-fueled and Carbon Nano Tubes catalyst were used in a study by Ramakrishnan et al., (2019). They found that biodiesel increased

NOx emission by about 4.8% in comparison to diesel fuel. It was also found that in the presence of carbon Nano Tubes catalyst, NOx emission decreased by 9.2% compared to neat biodiesel. Biodiesel decreased CO and HC by about 4.7% and 6.8%, respectively. However, the nanocarbon tube decreased them by 5.9 and 6.7%. In addition, CO_2 increased by about 6.6% with the use of biodiesel, which was increased in the presence of nanocatalyst. Smoke emission decreased by about 2.1% for biodiesel, while reaching 7.8% in the presence of nano.

Zhang et al., (2019) investigated the effect of low-level water addition on gas emission. The results showed that fuel-air mixing was enhanced because of micro-explosion. Also, results of this study indicated that low-level water addition in the biodiesel can decrease CO, NOx, and CO_2 emissions. The best water addition level in terms of gas emission was found at 4 wt% among WBE fuels. So, NOx, CO_2 and CO emissions were decreased at the maximum speed (1000 rpm). Therefore, the connection between water addition rate and emission features was not straightforward, and the balance between fuel and water addition was very significant.

Devarajan et al., (2019) carried out a survey of mahua biodiesel with copper oxide nanoparticle and found that BD100 caused lower levels of smoke, CO, HC, and NOx emissions. The presence of CuO nanoparticle in the biodiesel decreased NOx, CO, HC, and smoke emissions by around 3.9%, 4.9%, 5.6%, and 2.8% at the highest load situation in comparison to biodiesel. The increase of CuO nanoparticle in biodiesel resulted in more reduction of NOx, CO, HC, and smoke emissions.

Raju et al., (2019) studied the impact of exhaust gas recirculation, which operated by tamarind seed methyl ester. The biodiesel blend resulted in higher NOx emissions at all operating situations. The results showed that a biodiesel with 20% exhaust gas recirculation rate decreased NOx by about 52.69% and 45.67% compared with biodiesel and diesel fuels. Though, there was a minor decrease in the brake thermal efficiency.

Devarajan et al., (2019) worked on the effect of butanol, as an oxygenated additive to Calophyllum inophyllum, on CO, smoke, NOx, and HC emissions. Results showed lower HC, smoke, NOx and CO emissions in case of fuelling with both butanol blends in comparison to neat Punnai biodiesel.

Carbon Cycle

Carbon is the backbone of life on Earth. It is the foremost component of biological compounds and many minerals, such as limestone. It largely exists in the air as carbon dioxide. During the process of photosynthesis, carbon dioxide is converted into organic molecules by plants. Animals and humans use organic molecules through the food chains. The organic carbon is then converted back into carbon dioxide gas by cellular respiration. This life cycle is called carbon cycle.

As biodiesel is only produced from the oil contained in the plant seed and as the growth of plants needs CO_2, the production process of biodiesels is an important part of the carbon cycle, which dramatically reduces CO_2.

Net Energy Balance

Environmental advantages of biofuel are not enough to make it a viable substitute for fossil fuel. Biodiesels should be economical and producible in sufficient quantities in order to meet energy demands. Each step of the biodiesel production, from crop growth to biodiesel conversion, requires energy. Energy is needed for growing varietal or hybrid seed planted to generate the crop, powering farm machinery, producing pesticides and fertilizers, and sustaining farmers. It also needs energy for transporting the crop to biofuel production facilities and for their operations. Biodiesels are considered an energy content equal to their available energy upon combustion. So the net energy gained from biodiesel must be greater than the energy sources employed for its production. Jason's research shows that both corn grain ethanol and soybean biodiesel production result in positive Net Energy Balance (NEBs). It means that biofuel energy content exceeds fossil fuel energy inputs. His results also reveal that corn grain ethanol yields 25% more energy compared to the energy invested in its production. It is due to the high energy input needed to generate corn and convert it into ethanol; whereas, soybean biodiesel supplies 93% more energy than is needed for its production. For instant, the required energy for conversion of biomass to biofuel is much more for soybean biodiesel than corn grain ethanol. Because soybeans create long-chain triglycerides that are easily expressed from the seed, whereas corn starches must pass through enzymatic conversion into sugars, yeast fermentation to alcohol, and distillation to produce ethanol. These costs can be improved by employing agricultural residue or low-input biomass as corn Stover in lieu of fossil fuel energy in the biofuel conversion process (Hill et al., 2006).

BIODIESEL GENERATIONS

Different generations of biodiesel have been studied to date. Biodiesel typically falls into four different generations. Table 1 lists the oils used in the first generation to date.

The first generation of biodiesel consisted of edible oils. This generation has had a huge effect on food security and global food markets. Although there have been numerous reports on the impact of first-generation biodiesel use on world food security, there has been some exaggerations. However, due to the need for more markets for biodiesel production, it can have a direct impact on their prices in the long run. Other problems of this generation are environmental impacts. This is because with increased global demand for such a fuel, there is a growing need for commercial land, which leads to cultivation of more lands that has a direct impact on deforestation. Another environmental problem of the first generation is the increase of water consumption, which causes demand-generating countries face problems of underground water scarcity over time [22, 23].

Table 1. A list of the first generation of biofuels

First generation of biofuel					
Coconut oil (Coconut pulm, Copra oil)	Rapeseed oil	Cashew oil	Pistachio oil	Bitter ground oil (Momordica charantia)	Blackcurrant seed oil (Ribes nigrum)	False flax oil (Camelina sativa)
Corn oil (Maize oil)	Safflower oil	Hazelnut oil	Walnut oil	Egusi seed oil (Cucumeropsis mannii nandin)	Borage seed oil (Borago officinalis)	Grape seed oil
Canola oil	Sesame oil	Macadamia oil	Lemon oil	Pumpkin seed oil	Evening promise oil (Oenothera biennis)	Hemp oil
Cotton seed oil	Soybean oil	Mongongo nut oil (Manketti oil, Schinziophyton rautanenii)	Orange oil	Watermelon seed oil (Citrullus vulgaris)	Flaxseed oil (linseed oil, Linum usitatissimum)	Kapok seed oil (Ceiba pentandra)
Olive oil	Sunflower oil	Cashew oil	Okra seed oil (Abelmoschus esculentus)	Açai oil (Euterpe oleracea)	Apricot oil	Kenaf seed oil (Hibiscus cannabinus)
Palm oil	Almond oil	Meadowfoam seed oil	Papaya seed oil	Black seed oil (Nigella sativa)	Apple seed oil	Marula oil (Sclerocarya birrea)
Peanut oil (Arachis hypogea, Ground nut oil)	Beech nut oil (Fagus sylvatica)	Mustard oil	Rice bran oil	Bitter ground oil (Momordica charantia)	Avocado oil	

Table 1. (Continued)

First generation of biofuel

Pequi oil (*Caryocar brasiliense*)	Brazilnut oil (*Bertholletia excelsa*)	Niger seed oil (Asteraceae, *Guizotia abyssinica*)	Pomegranate seed oil (*Punica granatum*)	Taramira oil (*Eruca sativa*)	Tobacco seed oil (*Nicotiana tabacum*)
Sacha inchi oil	Colza oil (*Brassica rapa*)	Tea seed oil (*Camellia oil*)	Castor oil	Cocklebur oil (Xanthium)	Poppyseed oil
Sapote oil	Radish oil	Thistle oil (*Silybum marianum*)	Babassu oil (*Attalea speciousa*)	Coriander seed oil	Pracaxi oil (*Pentaclethra macroloba*)
Shea butter	Tigernut oil (Nut-sedge oil, *Cyperus esculentus*)	Tomato seed oil	Ben oil (*Moringa oleifera*)	Date seed oil	Dika oil (*Irvingia gabonensis*)

Table 2. A lists of the second generation of biofuels

Second generation of biofuel			
Tung oil	Stillingia oil (Chinese vegetable tallow oil, *Sapium sebiferum*)	Candlenut oil (Kukui nut oil)	Honesty oil (*Lunaria annua*)
Jatropha oil	Artichoke oil	Chaulmoogra oil (*Hydnocarpus wightiana*)	Mango oil
Jojoba oil (*Simmondsia chinensis*)	*Astrocaryum murumuru* butter	Crambe oil (*Crambe abyssinica*)	Neem oil (*Azadirachta indica*)
Nahor oil (*Mesua ferrea*)	Balanos oil (*Balanites aegyptiaca*)	Croton oil (tiglium oil, *Croton tiglium*)	-
Paradise oil (*Simarouba glauca*)	Brucea javanica oil	Cuphea oil	-
Pongamia oil (Karanja oil, Honge oil, *Millettia pinnata*)	Buriti oil (*Mauritia flexuosa*)	Cupuaçu butter	-
Tall oil	*Cebera odollam* (sea mango oil)	greenseed oil	-
Tamanu oil (Polanga oil, Foraha oil, *Calophyllum tacamahaca*)	nagchampa	*Aphanamixis polystachya*	-
Tucumã butter (*Astrocaryum vulgare*)	*Croton megalocarpus*	Rubber seed oil (*Hevea brasiliensis*)	-
Sapindus mukorossi (soapnut oil)	Patchouli oil (*Pogostemon cablin*)	Sea buckthorn oil (*Hippophae rhamnoides*)	-
Nerium oleander (*Thevetia peruviana*)	*Sterculia foetida*	Passion fruit oil (*Passiflora edulis*)	-

The second-generation oils, known as non-edible oils, are listed in Table 2 above.

The second generation is a suitable option for biodiesel production in order to reduce dependence on first generation oils. This generation of biodiesel has many advantages over the first generation. Most importantly, this generation will have no impact on food chain and will not increase their prices. This generation of biodiesel contains no toxic compounds that are harmful to humans. They are more environmentally friendly and more effective compared to the first generation of biodiesel. One of the greatest important benefits of this generation is the ability to grow in inferior lands, where the first generation is unable to grow. They also have valuable by-products, which can be burned for heat and power or chemical processes. One of the problems of this generation is the lack of sufficient volume for large scale production and growth in specific regions of the world. This will decline the amount of production and create a monopoly for specific countries.

Table 3. A list of the third generation of biofuels

Third generation of biofuel		
Ankistrodesmus braunii and *Nannochloropsis*	*Spirulina platensis*	inedible animal tallow
Auxenochorella protothecoides	Animal fat	Lard
Chlorella protothecoides	Animal fat traps	Mutton fat
Chlorella variabilis	Beef tallow	Poultry fat
Chlorella vulgaris	*Camelus dromedaries* fat (Camel fat)	Waste cooking oil
Euglena sanguinea	Chicken fat	Waste fat oil
Heterotrophic microalgae (Sugar plant)	Fish oil	Waste fried oil
Melanothamnus afaqhusainii	Fleshing oil	Waste frying palm oil
Pond water algae	Trout oil	Waste mixed vegetable oil
Schizochytrium mangrovei	Larvae grease (housefly)	Waste sunflower oil
Spirulina	Sludge pyrolysis oil	Neem seed pyrolysis oil

The third generation of biofuels is shown in Table 3. Many articles have been reported today about the impact of this generation that have different advantages and disadvantages. One of the most important benefits of this generation is that they have no impact on food chain. Also, this generation does not require arable cultivation land like the previous two generations. The crops of this generation can be grown under different conditions, solving the problem of monopoly. High growth capability and useful by-products that can be employed in other industries are other benefits of this generation. Also, algae has problems in this generation. Algae, even when grown in waste water, needs large amounts of water, nitrogen and phosphorus to grow. So, the production of fertilizer to meet the requirements of algae applied to produce biofuel would generate more greenhouse gas emissions. The price of algae-based biofuel is greater than fuel generated from other sources. This means that the large-scale implementation of algae to generate biofuel will not happen for the long time, if at all. In fact, after investing $600 million USD in research and development of algae, Exxon Mobil found that algae-based biofuels will not be viable for at least 25 years. In addition, this is firmly economical and does not consider the environmental impacts that have yet to be solved. A minor concern regarding algae is that its biofuel tends to be less stable compared to biodiesel produced from other sources, which is due to the fact that the oil found in algae tends to be extremely unsaturated. Unsaturated oils are more volatile, mostly at high temperatures, and therefore, more prone to degradation.

Biodiesel is mostly produced in several countries. Brazil and USA about 87% of biofuel of the world. Biodiesel production is around 1,890 thousand barrels/day in the world. The USA produced about 1,047 thousand barrels/day in 2018, contributing to 55.4% of biofuel production. The US mainly applied corn as the principal feedstock for generating ethanol fuel and soybeans for biodiesel production. Brazil generated 693.2 thousand barrels/day and took second place in generating biofuel. The South America held 26.5% share of the world's total biofuel production in 2018. In addition, Brazil was the second-ranked producer of biogasoline with a 31.5% share, equal to about 595.35 thousand barrels/day. The Latin

America mainly uses sugar cane for generating ethanol fuel and soybeans for biodiesel production. Germany is the third country producing 75.8 thousand barrels/day in 2018, accounting for 2.9% of global biofuel production capacity in 2018. The German Association of Biodiesel Producers (Verband der Deutschen Biokraftstoffindustrie, VDB) claimed that they generated 3.2 million tons of biodiesel in 2018. This country used cooking oil and rapeseed as raw materials for producing biofuel. Argentina is the next country, which produced the highest amount of biofuel by about 70.6 thousand barrels/day. This country supplied about 2.7% of the biofuel production in the market. They used nineteen bioethanol plants for biofuel production. Finally, China produced 68 thousand barrels/day as the fifth largest biofuel producer, accounting for 2.6% of total biofuel production.

Conclusion

Energy is essential for preserving economic growth and development. Globally, the largest part of energy consumption is allocated to the transportation sector, after the industrial sector. Almost all fossil fuel energy consumed in the transportation sector is produced from oil (97.6%). Since fossil fuel resources are decreasing day by day, there is a need to find out alternative fuels that are not only able to fulfill the energy demand of the world, but also help reduce global warming and environmental issues caused by fossil fuels' burning. Biodiesel is such a fuel that has gained lots of attention in recent decades. Biodiesel has several distinct advantages in comparison to petroleum fuels, such as carbon cycle, biodegradability, higher flash point, excellent lubricity, and non-toxicity. One of the disadvantages of biodiesel is its inherent higher price, which is compensated by legislative and regulatory incentives or subsidies in numerous countries. Commonly, it is produced through a transesterification reaction between a type of alcohol and oil in the presence of an acidic or a basic catalyst. This method of production is regarded as the best method from among other approaches (pyrolysis, dilution with hydrocarbons blending, microemulsion) due to its low cost and simplicity. There is a wide range of available feedstock for biodiesel production, such as vegetable oils (edible oils like rapeseed or canola and non-edible oils like jatropha or even microalgae), waste cooking oils, and animal fats. Thus, choosing the best feedstock is essential to lower the overall cost of biodiesel, since more than 75% of biodiesel production is related to feedstock.

ACKNOWLEDGMENTS

This research did not receive any specific grant from funding agencies in the public, commercial, or not-for-profit sectors.

REFERENCES

Abdullah, Bawadi, Syed Anuar Faua'ad Syed Muhammad, Zahra Shokravi, Shahrul Ismail, Khairul Anuar Kassim, Azmi Nik Mahmood, and Md Maniruzzaman A Aziz. "Fourth Generation Biofuel: A Review on Risks and Mitigation Strategies." *Renewable and Sustainable Energy Reviews* 107 (2019): 37-50.

Ahmad, AL, NH Mat Yasin, CJC Derek, and JK Lim. "Microalgae as a Sustainable Energy Source for Biodiesel Production: A Review." *Renewable and Sustainable Energy Reviews* 15, no. 1 (2011): 584-93.

Alptekin, Ertan. "Emission, Injection and Combustion Characteristics of Biodiesel and Oxygenated Fuel Blends in a Common Rail Diesel Engine." *Energy* 119 (2017): 44-52.

An, H, WM Yang, SK Chou, and KJ Chua. "Combustion and Emissions Characteristics of Diesel Engine Fueled by Biodiesel at Partial Load Conditions." *Applied Energy* 99 (2012): 363-71.

An, H, WM Yang, A Maghbouli, J Li, SK Chou, and KJ Chua. "Performance, Combustion and Emission Characteristics of Biodiesel Derived from Waste Cooking Oils." *Applied energy* 112 (2013): 493-99.

Anand, K, RP Sharma, and Pramod S Mehta. "Experimental Investigations on Combustion, Performance and Emissions Characteristics of Neat Karanji Biodiesel and Its Methanol Blend in a Diesel Engine." *Biomass and bioenergy* 35, no. 1 (2011): 533-41.

Armaroli, Nicola, and Vincenzo Balzani. "The Future of Energy Supply: Challenges and Opportunities." *Angewandte Chemie International Edition* 46, no. 1-2 (2007): 52-66.

Arul Gnana Dhas, Anderson, Yuvarajan Devarajan, and Beemkumar Nagappan. "Analysis of Emission Reduction in Ethyne–Biodiesel-Aspirated Diesel Engine." *International Journal of Green Energy* 15, no. 7 (2018): 436-40.

Ashraful, AM, Haji Hassan Masjuki, MA Kalam, HK Rashedul, M Habibullah, MM Rashed, MH Mosarof, and A Arslan. "Impact of Edible and Non-Edible Biodiesel Fuel Properties and Engine Operation Condition on the Performance and Emission Characteristics of Unmodified Di Diesel Engine." *Biofuels* 7, no. 3 (2016): 219-32.

Atlas, Ronald M. "Petroleum Biodegradation and Oil Spill Bioremediation." *Marine Pollution Bulletin* 31, no. 4-12 (1995): 178-82.

Aydin, Hüseyin, and Cumali Ilkılıc. "Effect of Ethanol Blending with Biodiesel on Engine Performance and Exhaust Emissions in a Ci Engine." *Applied Thermal Engineering* 30, no. 10 (2010): 1199-204.

Babazadeh, Reza, Jafar Razmi, Mir Saman Pishvaee, and Masoud Rabbani. "A Sustainable Second-Generation Biodiesel Supply Chain Network Design Problem under Risk." *Omega* 66 (2017): 258-77.

Bakeas, Evangelos, Georgios Karavalakis, and Stamoulis Stournas. "Biodiesel Emissions Profile in Modern Diesel Vehicles. Part 1: Effect of Biodiesel Origin on the Criteria Emissions." *Science of the Total Environment* 409, no. 9 (2011): 1670-76.

Balan, KN, U Yashvanth, P Booma Devi, T Arvind, H Nelson, and Yuvarajan Devarajan. "Investigation on Emission Characteristics of Alcohol Biodiesel Blended Diesel Engine." *Energy Sources, Part A: Recovery, Utilization, and Environmental Effects* 41, no. 15 (2019): 1879-89.

Barabas, Istvan, Adrian Todoruţ, and Doru Băldean. "Performance and Emission Characteristics of an Ci Engine Fueled with Diesel–Biodiesel–Bioethanol Blends." *Fuel* 89, no. 12 (2010): 3827-32.

Bari, Saiful, CW Yu, and TH Lim. "Performance Deterioration and Durability Issues While Running a Diesel Engine with Crude Palm Oil." *Proceedings of the Institution of Mechanical Engineers, Part D: Journal of Automobile Engineering* 216, no. 9 (2002): 785-92.

Beer, Tom, T Grant, G Morgan, J Lapszewicz, P Anyon, J Edwards, P Nelson, H Watson, and D Williams. *Comparison of Transport Fuels: Final Report (Ev45a/2/F3c) to the Australian Greenhouse Office on the Stage 2 Study of Life-Cycle Emissions Analysis of Alternative Fuels for Heavy Vehicles.* (2001).

Behçet, Rasim. "Performance and Emission Study of Waste Anchovy Fish Biodiesel in a Diesel Engine." *Fuel Processing Technology* 92, no. 6 (2011): 1187-94.

Bhuiya, MMK, MG Rasul, Mohammad Masud Kamal Khan, Nanjappa Ashwath, Abul Kalam Azad, and MA Hazrat. "Second Generation Biodiesel: Potential Alternative to-Edible Oil-Derived Biodiesel." *Energy Procedia* 61 (2014): 1969-72.

Bolszo, CD, and VG McDonell. "Emissions Optimization of a Biodiesel Fired Gas Turbine." *Proceedings of the Combustion Institute* 32, no. 2 (2009): 2949-56.

Buyukkaya, Ekrem. "Effects of Biodiesel on a Di Diesel Engine Performance, Emission and Combustion Characteristics." *Fuel* 89, no. 10 (2010): 3099-105.

Can, Özer. "Combustion Characteristics, Performance and Exhaust Emissions of a Diesel Engine Fueled with a Waste Cooking Oil Biodiesel Mixture." *Energy Conversion and Management* 87 (2014): 676-86.

Can, Özer, Erkan Öztürk, Hamit Solmaz, Fatih Aksoy, Can Çinar, and H Serdar Yücesu. "Combined Effects of Soybean Biodiesel Fuel Addition and Egr Application on the Combustion and Exhaust Emissions in a Diesel Engine." *Applied thermal engineering* 95 (2016): 115-24.

Canakci, Mustufa, and J Van Gerpen. "Biodiesel Production from Oils and Fats with High Free Fatty Acids." *Transactions of the ASAE* 44, no. 6 (2001): 1429.

Chauhan, Bhupendra Singh, Naveen Kumar, and Haeng Muk Cho. "A Study on the Performance and Emission of a Diesel Engine Fueled with Jatropha Biodiesel Oil and Its Blends." *Energy* 37, no. 1 (2012): 616-22.

Cheikh, Kezrane, Awad Sary, Loubar Khaled, Liazid Abdelkrim, and Tazerout Mohand. "Experimental Assessment of Performance and Emissions Maps for Biodiesel Fueled Compression Ignition Engine." *Applied energy* 161 (2016): 320-29.

Choi, Seung-Hun, and Younhtaig Oh. "The Emission Effects by the Use of Biodiesel Fuel." *International Journal of Modern Physics B* 20, no. 25n27 (2006): 4481-86.

Chouhan, AP Singh, and AK Sarma. "Modern Heterogeneous Catalysts for Biodiesel Production: A Comprehensive Review." *Renewable and sustainable energy reviews* 15, no. 9 (2011): 4378-99.

Devarajan, Yuvarajan. "Experimental Evaluation of Combustion, Emission and Performance of Research Diesel Engine Fuelled Di-Methyl-Carbonate and Biodiesel Blends." *Atmospheric Pollution Research* 10, no. 3 (2019): 795-801.

Devarajan, Yuvarajan, Ravi kumar Jayabal, Devanathan Ragupathy, and Harish Venu. "Emissions Analysis on Second Generation Biodiesel." *Frontiers of Environmental Science & Engineering* 11, no. 1 (2017): 3.

Devarajan, Yuvarajan, Arulprakasajothi Mahalingam, Dinesh Babu Munuswamy, and T Arunkumar. "Combustion, Performance, and Emission Study of a Research Diesel Engine Fueled with Palm Oil Biodiesel and Its Additive." *Energy & fuels* 32, no. 8 (2018): 8447-52.

Devarajan, Yuvarajan, Dinesh Babu Munuswamy, and Arulprakasajothi Mahalingam. "Influence of Nano-Additive on Performance and Emission Characteristics of a Diesel Engine Running on Neat Neem Oil Biodiesel." *Environmental Science and Pollution Research* 25, no. 26 (2018): 26167-72.

Devarajan, Yuvarajan, Dinesh Babu Munuswamy, Beemkumar Nagappan, and Amith Kishore Pandian. "Performance, Combustion and Emission Analysis of Mustard Oil Biodiesel and Octanol Blends in Diesel Engine." *Heat and Mass Transfer* 54, no. 6 (2018): 1803-11.

Devarajan, Yuvarajan, Dineshbabu Munuswamy, Beemkumar Nagappan, and Ganesan Subbiah. "Experimental Assessment of Performance and Exhaust Emission Characteristics of a Diesel Engine Fuelled with Punnai Biodiesel/Butanol Fuel Blends." *Petroleum Science* (2019): 1-8.

Devarajan, Yuvarajan, Beemkumar Nagappan, and Ganesan Subbiah. "A Comprehensive Study on Emission and Performance Characteristics of a Diesel Engine Fueled with Nanoparticle-Blended Biodiesel." *Environmental Science and Pollution Research* 26, no. 11 (2019): 10662-72.

Dhana Raju, V, and PS Kishore. "Effect of Exhaust Gas Recirculation on Performance and Emission Characteristics of a Diesel Engine Fuelled with Tamarind Biodiesel." *International Journal of Ambient Energy* 40, no. 6 (2019): 624-33.

Di, Yage, CS Cheung, and Zuohua Huang. "Experimental Investigation on Regulated and Unregulated Emissions of a Diesel Engine Fueled with Ultra-Low Sulfur Diesel Fuel Blended with Biodiesel from Waste Cooking Oil." *Science of the total environment* 407, no. 2 (2009): 835-46.

Estill, Lyle. *Biodiesel Power: The Passion, the People, and the Politics of the Next Renewable Fuel.* New Society Publishers Gabriola Island, BC, 2005.

Fang, Zhen. *Biodiesel: Feedstocks, Production and Applications.* BoD–Books on Demand, 2012.

Fattah, IM Rizwanul, HH Masjuki, MA Kalam, M Mofijur, and MJ Abedin. "Effect of Antioxidant on the Performance and Emission Characteristics of a Diesel Engine Fueled with Palm Biodiesel Blends." *Energy Conversion and Management* 79 (2014): 265-72.

Fattah, IM Rizwanul, HH Masjuki, MA Kalam, MA Wakil, AM Ashraful, and S Ashraful Shahir. "Experimental Investigation of Performance and Regulated Emissions of a Diesel Engine with Calophyllum Inophyllum Biodiesel Blends Accompanied by Oxidation Inhibitors." *Energy Conversion and Management* 83 (2014): 232-40.

Ganapathy, T, RP Gakkhar, and K Murugesan. "Influence of Injection Timing on Performance, Combustion and Emission Characteristics of Jatropha Biodiesel Engine." *Applied energy* 88, no. 12 (2011): 4376-86.

Gharehghani, Ayatallah, Mostafa Mirsalim, and Reza Hosseini. "Effects of Waste Fish Oil Biodiesel on Diesel Engine Combustion Characteristics and Emission." *Renewable Energy* 101 (2017): 930-36.

Gnanasekaran, Sakthivel, N Saravanan, and M Ilangkumaran. "Influence of Injection Timing on Performance, Emission and Combustion Characteristics of a Di Diesel Engine Running on Fish Oil Biodiesel." *Energy* 116 (2016): 1218-29.

Gumus, M, and S Kasifoglu. "Performance and Emission Evaluation of a Compression Ignition Engine Using a Biodiesel (Apricot Seed Kernel Oil Methyl Ester) and Its Blends with Diesel Fuel." *Biomass and bioenergy* 34, no. 1 (2010): 134-39.

Gumus, Metin, Cenk Sayin, and Mustafa Canakci. "The Impact of Fuel Injection Pressure on the Exhaust Emissions of a Direct Injection Diesel Engine Fueled with Biodiesel–Diesel Fuel Blends." *Fuel* 95 (2012): 486-94.

Handling, Biodiesel, and Use Guide Fourth Edition. *Nrel*. TP-540-43672 Revised January (2009).

Hill, Jason, Erik Nelson, David Tilman, Stephen Polasky, and Douglas Tiffany. "Environmental, Economic, and Energetic Costs and Benefits of Biodiesel and Ethanol Biofuels." *Proceedings of the National Academy of sciences* 103, no. 30 (2006): 11206-10.

Hoseini, SS, G Najafi, B Ghobadian, R Mamat, MT Ebadi, and Talal Yusaf. "Novel Environmentally Friendly Fuel: The Effects of Nanographene Oxide Additives on the Performance and Emission Characteristics of Diesel Engines Fuelled with Ailanthus Altissima Biodiesel." *Renewable energy* 125 (2018): 283-94.

Hulwan, Dattatray Bapu, and Satishchandra V Joshi. "Performance, Emission and Combustion Characteristic of a Multicylinder Di Diesel Engine Running on Diesel–Ethanol–Biodiesel Blends of High Ethanol Content." *Applied Energy* 88, no. 12 (2011): 5042-55.

Ilgen, Oguzhan. "Dolomite as a Heterogeneous Catalyst for Transesterification of Canola Oil." *Fuel Processing Technology* 92, no. 3 (2011): 452-55.

Imdadul, HK, HH Masjuki, MA Kalam, NWM Zulkifli, Abdullah Alabdulkarem, MM Rashed, YH Teoh, and HG How. "Higher Alcohol–Biodiesel–Diesel Blends: An Approach for Improving the Performance, Emission, and Combustion of a Light-Duty Diesel Engine." *Energy Conversion and Management* 111 (2016): 174-85.

Joy, Nivin, Yuvarajan Devarajan, Beemkumar Nagappan, and A Anderson. "Exhaust Emission Study on Neat Biodiesel and Alcohol Blends Fueled Diesel Engine." *Energy Sources, Part A: Recovery, Utilization, and Environmental Effects* 40, no. 1 (2018): 115-19.

Joy, Nivin, Devarajan Yuvarajan, and Nagappan Beemkumar. "Performance Evaluation and Emission Characteristics of Biodiesel-Ignition Enhancer Blends Propelled in a Research Diesel Engine." *International journal of green energy* 16, no. 4 (2019): 277-83.

Kannan, GR, Ramasamy Karvembu, and R Anand. "Effect of Metal Based Additive on Performance Emission and Combustion Characteristics of Diesel Engine Fuelled with Biodiesel." *Applied energy* 88, no. 11 (2011): 3694-703.

Karabektas, Murat. "The Effects of Turbocharger on the Performance and Exhaust Emissions of a Diesel Engine Fuelled with Biodiesel." *Renewable Energy* 34, no. 4 (2009): 989-93.

Kim, Hwanam, and Byungchul Choi. "The Effect of Biodiesel and Bioethanol Blended Diesel Fuel on Nanoparticles and Exhaust Emissions from Crdi Diesel Engine." *Renewable energy* 35, no. 1 (2010): 157-63.

Knothe, Gerhard, Jürgen Krahl, and Jon Van Gerpen. *The Biodiesel Handbook*. Elsevier, 2015.

Krahl, Jürgen, Gerhard Knothe, Axel Munack, Yvonne Ruschel, Olaf Schröder, Ernst Hallier, Götz Westphal, and Jürgen Bünger. "Comparison of Exhaust Emissions and Their Mutagenicity from the Combustion of Biodiesel, Vegetable Oil, Gas-to-Liquid and Petrodiesel Fuels." *Fuel* 88, no. 6 (2009): 1064-69.

Leung, Dennis YC, Xuan Wu, and MKH Leung. "A Review on Biodiesel Production Using Catalyzed Transesterification." *Applied energy* 87, no. 4 (2010): 1083-95.

Li, Bowen, Yanfei Li, Haoye Liu, Fang Liu, Zhi Wang, and Jianxin Wang. "Combustion and Emission Characteristics of Diesel Engine Fueled with Biodiesel/Pode Blends." *Applied energy* 206 (2017): 425-31.

Li, Li, Jianxin Wang, Zhi Wang, and Jianhua Xiao. "Combustion and Emission Characteristics of Diesel Engine Fueled with Diesel/Biodiesel/Pentanol Fuel Blends." *Fuel* 156 (2015): 211-18.

Li, Shiwu, Yunpeng Wang, Shengwu Dong, Yang Chen, Fenghua Cao, Fang Chai, and Xiaohong Wang. "Biodiesel Production from Eruca Sativa Gars Vegetable Oil and Motor, Emissions Properties." *Renewable Energy* 34, no. 7 (2009): 1871-76.

Liaquat, AM, Haji Hassan Masjuki, MA Kalam, IM Rizwanul Fattah, MA Hazrat, Mahendra Varman, M Mofijur, and M Shahabuddin. "Effect of Coconut Biodiesel Blended Fuels on Engine Performance and Emission Characteristics." *Procedia Engineering* 56 (2013): 583-90.

Lin, Lin, Dong Ying, Sumpun Chaitep, and Saritporn Vittayapadung. "Biodiesel Production from Crude Rice Bran Oil and Properties as Fuel." *Applied Energy* 86, no. 5 (2009): 681-88.

Ma, Fangrui, and Milford A Hanna. "Biodiesel Production: A Review." *Bioresource technology* 70, no. 1 (1999): 1-15.

Macor, A, and P Pavanello. "Performance and Emissions of Biodiesel in a Boiler for Residential Heating." *Energy* 34, no. 12 (2009): 2025-32.

Mahalingam, Arulprakasajothi, Yuvarajan Devarajan, Santhanakrishnan Radhakrishnan, Suresh Vellaiyan, and Beemkumar Nagappan. "Emissions Analysis on Mahua Oil Biodiesel and Higher Alcohol Blends in Diesel Engine." *Alexandria Engineering Journal* 57, no. 4 (2018): 2627-31.

Mahalingam, Arulprakasajothi, Dinesh Babu Munuswamy, Yuvarajan Devarajan, and Santhanakrishnan Radhakrishnan. "Emission and Performance Analysis on the Effect of Exhaust Gas Recirculation in Alcohol-Biodiesel Aspirated Research Diesel Engine." *Environmental Science and Pollution Research* 25, no. 13 (2018): 12641-47.

Man, XJ, CS Cheung, Z Ning, L Wei, and ZH Huang. "Influence of Engine Load and Speed on Regulated and Unregulated Emissions of a Diesel Engine Fueled with Diesel Fuel Blended with Waste Cooking Oil Biodiesel." *Fuel* 180 (2016): 41-49.

Mazza, Patrick, and Roel Hammerschlag. "Wind-to-Wheel Energy Assessment." Paper presented at the *Proceedings of the European Fuel Cell Forum*, http://www. efcf. com/reports E, 2005.

Mirzajanzadeh, Mehrdad, Meisam Tabatabaei, Mehdi Ardjmand, Alimorad Rashidi, Barat Ghobadian, Mohammad Barkhi, and Mohammad Pazouki. "A Novel Soluble Nano-Catalysts in Diesel–Biodiesel Fuel Blends to Improve Diesel Engines Performance and Reduce Exhaust Emissions." *Fuel* 139 (2015): 374-82.

Mohsin, R, ZA Majid, AH Shihnan, NS Nasri, and Z Sharer. "Effect of Biodiesel Blends on Engine Performance and Exhaust Emission for Diesel Dual Fuel Engine." *Energy conversion and management* 88 (2014): 821-28.

Mueller, Charles J, André L Boehman, and Glen C Martin. "An Experimental Investigation of the Origin of Increased Nox Emissions When Fueling a Heavy-Duty Compression-Ignition Engine with Soy Biodiesel." *SAE International Journal of Fuels and Lubricants* 2, no. 1 (2009): 789-816.

Muralidharan, K, D Vasudevan, and KN Sheeba. "Performance, Emission and Combustion Characteristics of Biodiesel Fuelled Variable Compression Ratio Engine." *Energy* 36, no. 8 (2011): 5385-93.

Nabi, Md Nurun, SM Najmul Hoque, and Md Shamim Akhter. "Karanja (Pongamia Pinnata) Biodiesel Production in Bangladesh, Characterization of Karanja Biodiesel and Its Effect on Diesel Emissions." *Fuel processing technology* 90, no. 9 (2009): 1080-86.

Nabi, Md Nurun, Md Mustafizur Rahman, and Md Shamim Akhter. "Biodiesel from Cotton Seed Oil and Its Effect on Engine Performance and Exhaust Emissions." *Applied thermal engineering* 29, no. 11-12 (2009): 2265-70.

Nabi, Md Nurun, Ali Zare, Farhad M Hossain, Zoran D Ristovski, and Richard J Brown. "Reductions in Diesel Emissions Including Pm and Pn Emissions with Diesel-Biodiesel Blends." *Journal of cleaner production* 166 (2017): 860-68.

Najafi, G. "Diesel Engine Combustion Characteristics Using Nano-Particles in Biodiesel-Diesel Blends." *Fuel* 212 (2018): 668-78.

Nalgundwar, Ankur, Biswajit Paul, and Sunil Kumar Sharma. "Comparison of Performance and Emissions Characteristics of Di Ci Engine Fueled with Dual Biodiesel Blends of Palm and Jatropha." *Fuel* 173 (2016): 172-79.

Nithya, S, S Manigandan, P Gunasekar, J Devipriya, and WSR Saravanan. "The Effect of Engine Emission on Canola Biodiesel Blends with Tio2." *International Journal of Ambient Energy* 40, no. 8 (2019): 838-41.

Omidvarborna, Hamid, Ashok Kumar, and Dong-Shik Kim. "Nox Emissions from Low-Temperature Combustion of Biodiesel Made of Various Feedstocks and Blends." *Fuel Processing Technology* 140 (2015): 113-18.

Özener, Orkun, Levent Yüksek, Alp Tekin Ergenç, and Muammer Özkan. "Effects of Soybean Biodiesel on a Di Diesel Engine Performance, Emission and Combustion Characteristics." *Fuel* 115 (2014): 875-83.

Öztürk, Erkan. "Performance, Emissions, Combustion and Injection Characteristics of a Diesel Engine Fuelled with Canola Oil–Hazelnut Soapstock Biodiesel Mixture." *Fuel Processing Technology* 129 (2015): 183-91.

Pandian, Amith Kishore, Dinesh Babu Munuswamy, Santhanakrishnan Radhakrishana, Ramesh Bapu Bathey Ramakrishnan, Beemkumar Nagappan, and Yuvarajan Devarajan. "Influence of an Oxygenated Additive on Emission of an Engine Fueled with Neat Biodiesel." *Petroleum Science* 14, no. 4 (2017): 791-97.

Pandian, Amith Kishore, Ramesh Bapu Bathey Ramakrishnan, and Yuvarajan Devarajan. "Emission Analysis on the Effect of Nanoparticles on Neat Biodiesel in Unmodified Diesel Engine." *Environmental Science and Pollution Research* 24, no. 29 (2017): 23273-78.

Pinto, Angelo C, Lilian LN Guarieiro, Michelle JC Rezende, Núbia M Ribeiro, Ednildo A Torres, Wilson A Lopes, Pedro A de P Pereira, and Jailson B de Andrade. "Biodiesel: An Overview." *Journal of the Brazilian Chemical Society* 16, no. 6B (2005): 1313-30.

Pinzi, Sara, Paul Rounce, José M Herreros, Athanasios Tsolakis, and M Pilar Dorado. "The Effect of Biodiesel Fatty Acid Composition on Combustion and Diesel Engine Exhaust Emissions." *Fuel* 104 (2013): 170-82.

Prabu, A, and RB Anand. "Emission Control Strategy by Adding Alumina and Cerium Oxide Nano Particle in Biodiesel." *Journal of the Energy Institute* 89, no. 3 (2016): 366-72.

Qi, DH, H Chen, LM Geng, and YZ Bian. "Effect of Diethyl Ether and Ethanol Additives on the Combustion and Emission Characteristics of Biodiesel-Diesel Blended Fuel Engine." *Renewable energy* 36, no. 4 (2011): 1252-58.

Radhakrishnan, Santhanakrishnan, Yuvarajan Devarajan, Arulprakasajothi Mahalingam, and Beemkumar Nagappan. "Emissions Analysis on Diesel Engine Fueled with Palm Oil Biodiesel and Pentanol Blends." *Journal of Oil Palm Research* 29, no. 3 (2017): 380-86.

Radhakrishnan, Santhanakrishnan, Dinesh Babu Munuswamy, Yuvarajan Devarajan, and Arulprakasajothi Mahalingam. "Effect of Nanoparticle on Emission and Performance Characteristics of a Diesel Engine Fueled with Cashew Nut Shell Biodiesel." *Energy Sources, Part A: Recovery, Utilization, and Environmental Effects* 40, no. 20 (2018): 2485-93.

Rajak, Upendra, Prerana Nashine, and Tikendra Nath Verma. "Performance Analysis and Exhaust Emissions of Aegle Methyl Ester Operated Compression Ignition Engine." *Thermal Science and Engineering Progress* (2019).

Ramakrishnan, Ganaram, Purushothaman Krishnan, Sivasubramanian Rathinam, and Yuvarajan Devarajan. "Role of Nano-Additive Blended Biodiesel on Emission Characteristics of the Research Diesel Engine." *International Journal of Green Energy* 16, no. 6 (2019): 435-41.

Rathinam, Sivasubramanian, Sajin Justin Abraham Baby, and Yuvarajan Devarajan. "Influence of Water on Exhaust Emissions on Unmodified Diesel Engine Propelled with Biodiesel." *Energy Sources, Part A: Recovery, Utilization, and Environmental Effects* 40, no. 21 (2018): 2511-17.

Rounce, P, A Tsolakis, J Rodríguez-Fernández, APE York, RF Cracknell, and RH Clark. *Diesel Engine Performance and Emissions When First Generation Meets Next Generation Biodiesel.* SAE Technical Paper (2009).

Ryu, Kyunghyun. "The Characteristics of Performance and Exhaust Emissions of a Diesel Engine Using a Biodiesel with Antioxidants." *Bioresource Technology* 101, no. 1 (2010): S78-S82.

Sahoo, PK, LM Das, MKG Babu, P Arora, VP Singh, NR Kumar, and TS Varyani. "Comparative Evaluation of Performance and Emission Characteristics of Jatropha, Karanja and Polanga Based Biodiesel as Fuel in a Tractor Engine." *Fuel* 88, no. 9 (2009): 1698-707.

Saravankumar, PT, V Suresh, V Vijayan, and A Godwin Antony. "Ecological Effect of Corn Oil Biofuel with Sio2 Nano-Additives." *Energy Sources, Part A: Recovery, Utilization, and Environmental Effects* (2019): 1-8.

Sarin, Amit. *Biodiesel: Production and Properties.* Royal Society of Chemistry, 2012.

Sdrula, Nicolae. "A Study Using Classical or Membrane Separation in the Biodiesel Process." *Desalination* 250, no. 3 (2010): 1070-72.

Semwal, Surbhi, Ajay K Arora, Rajendra P Badoni, and Deepak K Tuli. "Biodiesel Production Using Heterogeneous Catalysts." *Bioresource technology* 102, no. 3 (2011): 2151-61.

Shahid, Ejaz M, and Younis Jamal. "Production of Biodiesel: A Technical Review." *Renewable and Sustainable Energy Reviews* 15, no. 9 (2011): 4732-45.

Sharma, YC, and B Singh. "Development of Biodiesel: Current Scenario." *Renewable and Sustainable Energy Reviews* 13, no. 6-7 (2009): 1646-51.

Sharp, Christopher A. *Emissions and Lubricity Evaluation of Rapeseed Derived Biodiesel Fuels*. Southwest Research Institute, 1996.

Siva, R, Dinesh Babu Munuswamy, and Yuvarajan Devarajan. "Emission and Performance Study Emulsified Orange Peel Oil Biodiesel in an Aspirated Research Engine." *Petroleum Science* 16, no. 1 (2019): 180-86.

Sreenath, Jishnu, and Anand Pai. "Biodiesel: A Review on Next Generation Fuels." *J. Adv. Res. Fluid Mech. Therm. Sci* 43 (2018): 58-66.

Tan, Pi-qiang, Zhi-yuan Hu, Di-ming Lou, and Zhi-jun Li. "Exhaust Emissions from a Light-Duty Diesel Engine with Jatropha Biodiesel Fuel." *Energy* 39, no. 1 (2012): 356-62.

Tan, Yie Hua, Mohammad Omar Abdullah, Cirilo Nolasco-Hipolito, Nur Syuhada Ahmad Zauzi, and Georgie Wong Abdullah. "Engine Performance and Emissions Characteristics of a Diesel Engine Fueled with Diesel-Biodiesel-Bioethanol Emulsions." *Energy Conversion and Management* 132 (2017): 54-64.

Trombettoni, Valeria, Daniela Lanari, Pepijn Prinsen, Rafael Luque, Assunta Marrocchi, and Luigi Vaccaro. "Recent Advances in Sulfonated Resin Catalysts for Efficient Biodiesel and Bio-Derived Additives Production." *Progress in Energy and Combustion Science* 65 (2018): 136-62.

Tsai, Jen-Hsiung, Shui-Jen Chen, Kuo-Lin Huang, Yuan-Chung Lin, Wen-Jhy Lee, Chih-Chung Lin, and Wen-Yinn Lin. "Pm, Carbon, and Pah Emissions from a Diesel Generator Fuelled with Soy-Biodiesel Blends." *Journal of hazardous materials* 179, no. 1-3 (2010): 237-43.

Uyumaz, Ahmet. "Combustion, Performance and Emission Characteristics of a Di Diesel Engine Fueled with Mustard Oil Biodiesel Fuel Blends at Different Engine Loads." *Fuel* 212 (2018): 256-67.

Vedaraman, N, Sukumar Puhan, G Nagarajan, and KC Velappan. "Preparation of Palm Oil Biodiesel and Effect of Various Additives on

Nox Emission Reduction in B20: An Experimental Study." *International Journal of Green Energy* 8, no. 3 (2011): 383-97.

Velmurugan, Ramanathan, Jaikumar Mayakrishnan, S Induja, Selvakumar Raja, Sasikumar Nandagopal, and Ravishankar Sathyamurthy. "Comprehensive Study on the Effect of Cuo Nano Fluids Prepared Using One-Step Chemical Synthesis Method on the Behavior of Waste Cooking Oil Biodiesel in Compression Ignition Engine." *Journal of Thermal Science and Engineering Applications* 11, no. 4 (2019): 041003.

Venu, Harish, Lingesan Subramani, and V Dhana Raju. "Emission Reduction in a Di Diesel Engine Using Exhaust Gas Recirculation (Egr) of Palm Biodiesel Blended with Tio2 Nano Additives." *Renewable Energy* (2019).

Wahlen, Bradley D, Michael R Morgan, Alex T McCurdy, Robert M Willis, Michael D Morgan, Daniel J Dye, Bruce Bugbee, Byard D Wood, and Lance C Seefeldt. "Biodiesel from Microalgae, Yeast, and Bacteria: Engine Performance and Exhaust Emissions." *Energy & Fuels* 27, no. 1 (2012): 220-28.

Yan, Fang, Zhenhong Yuan, Pengmei Lu, Wen Luo, Lingmei Yang, and Li Deng. "Fe–Zn Double-Metal Cyanide Complexes Catalyzed Biodiesel Production from High-Acid-Value Oil." *Renewable energy* 36, no. 7 (2011): 2026-31.

Yilmaz, Nadir, Francisco M Vigil, Kyle Benalil, Stephen M Davis, and Antonio Calva. "Effect of Biodiesel–Butanol Fuel Blends on Emissions and Performance Characteristics of a Diesel Engine." *Fuel* 135 (2014): 46-50.

Yoon, Seung Hyun, Hyun Kyu Suh, and Chang Sik Lee. "Effect of Spray and Egr Rate on the Combustion and Emission Characteristics of Biodiesel Fuel in a Compression Ignition Engine." *Energy & Fuels* 23, no. 3 (2009): 1486-93.

Zhang, Xiulin, Charles Peterson, Daryl Reece, R Haws, and G Möller. "Biodegradability of Biodiesel in the Aquatic Environment." *Transactions of the ASAE* 41, no. 5 (1998): 1423.

Zhang, Zhiqing, E Jiaqiang, Jingwei Chen, Hao Zhu, Xiaohuan Zhao, Dandan Han, Wei Zuo, et al., "Effects of Low-Level Water Addition on Spray, Combustion and Emission Characteristics of a Medium Speed Diesel Engine Fueled with Biodiesel Fuel." *Fuel* 239 (2019): 245-62.

Zhang, Zhiqing, E Jiaqiang, Yuanwang Deng, MinhHieu Pham, Wei Zuo, Qingguo Peng, and Zibin Yin. "Effects of Fatty Acid Methyl Esters Proportion on Combustion and Emission Characteristics of a Biodiesel Fueled Marine Diesel Engine." *Energy Conversion and Management* 159 (2018): 244-53.

Zheng, Zunqing, XiaoFeng Wang, Xiaofan Zhong, Bin Hu, Haifeng Liu, and Mingfa Yao. "Experimental Study on the Combustion and Emissions Fueling Biodiesel/N-Butanol, Biodiesel/Ethanol and Biodiesel/2, 5-Dimethylfuran on a Diesel Engine." *Energy* 115 (2016): 539-49.

Zhu, Lei, CS Cheung, WG Zhang, and Zhen Huang. "Emissions Characteristics of a Diesel Engine Operating on Biodiesel and Biodiesel Blended with Ethanol and Methanol." *Science of the Total Environment* 408, no. 4 (2010): 914-21.

In: Biofuels
Editor: George R. Carey

ISBN: 978-1-53617-721-3
© 2020 Nova Science Publishers, Inc.

Chapter 4

NON-CONVENTIONAL YEASTS WITH POTENTIAL FOR PRODUCTION OF SECOND-GENERATION ETHANOL

Katharina O. Barros[1], Angela M. Garcia-Acero[1,2] and Carlos A. Rosa[1],*

[1]Department of Microbiology, Institute of Biological Sciences, Federal University of Minas Gerais, Belo Horizonte-MG, Brazil
[2]Departament of Chemical and Environmental Engineering, Faculty of Engineering, National University of Colombia, Bogotá, Colombia

ABSTRACT

Second-generation (2G) ethanol production is dependent on the efficient conversion of the carbohydrates from the cellulosic and hemicellulosic fractions of lignocellulose, primarily D-glucose and D-xylose. Although, pentose is present in high levels in this substrate, *Saccharomyces cerevisiae*, the most widely used microorganism for industrial alcoholic fermentation, cannot metabolize this sugar. Non-conventional yeast species that are able to ferment xylose have been found in habitats such as rotting wood, tree bark, and leaves, and some of

---

* Corresponding Author's E-mail: carlrosa@icb.ufmg.br.

them are even associated with insects. *Pachysolen tannophilus* was the first species described to convert xylose directly to ethanol. Thereafter, other xylose-fermenting yeasts able to ferment this pentose to ethanol were isolated and identified. Among these, the ones that stand out are the species of the clades *Scheffersomyces* and *Spathaspora* as *Sc. stipitis*, *Sc. shehatae*, *Sc. lignosus*, *Sp. passalidarum*, *Sp. arborariae*, *Sp. gorwiae*, *Sp. hagerdaliae*, and *Sp. piracicabensis*. *Spathaspora passalidarum* was isolated from the gut of host beetles and rotting wood, and it is considered the most prominent species for the transformation of xylose to ethanol. Several yeasts are also capable of producing xylanolytic enzymes that degrade xylan, the major polysaccharide in the hemicellulose structure. The production of hydrolytic enzymes for the enzymatic hydrolysis process is reported by optimizing lignocellulose degradation and increasing the yield of simple sugars. Some basidiomycetous and ascomycetous yeasts produce xylanases and β-xylosidases from substrates such as xylan and D-xylose. The yeast-like fungus *Aurobasidium pullulans* is known to be a source of xylanolytic enzymes with high specificity. It can be used for simultaneous saccharification and fermentation with xylose-fermenting yeasts to improve the production of lignocellulosic ethanol. Xylanolytic yeasts also include species of the clades *Sugiyamaella* (*Su. xylanicola*, *Su. lignohabitans*, *Su. valenteae*, and *Su. smithiae*), *Lodderomyces*/*Candida albicans* (*C. tropicalis*), and *Clavispora*/*Candida* (*C. intermedia*); and the genera *Scheffersomyces* (*Sc. shehatae* and *Sc. sipitis*), *Naganishia* (*N. diffluens*), *Kwoniella* (*Kw. heveanensis*), and *Papiliotrema* (*Pa. laurentii*). Yeasts with the ability to ferment sugars of lignocellulose and/or produce enzymes that act on this substrate have great potential for being applied to the production of 2G ethanol.

Keywords: second-generation ethanol, lignocellulose, xylose-fermenting yeasts, xylanolytic enzymes

1. INTRODUCTION

The need to diversify energy origins in the world to transition into the production of fuels that are not derived from fossil resources is of great interest and comes with great technical challenges for feasible and scalable industrial production using different renewable resources. Fuels generated from biomass have been one of the solutions that have emerged as promising energy sources due to the cost-effectiveness of their feedstock;

biomass makes up more than 10% of the global energy supply (Wei et al. 2017).

In liquid biofuels, ethanol has immense potential as a substitute fuel for transportation (Arora et al., 2019). In the last decade, world ethanol production increased by 39% (Figure 1) and in the last year, the United States produced 16.100 million gallons of ethanol, Brazil generated 7.950 million gallons and the European Union produced 1.430 million gallons, being the largest producers in the world supporting the alcohol production from crops of feed interest such as corn, sugarcane, and wheat (also sugar beet), respectively (Zhang et al., 2018).

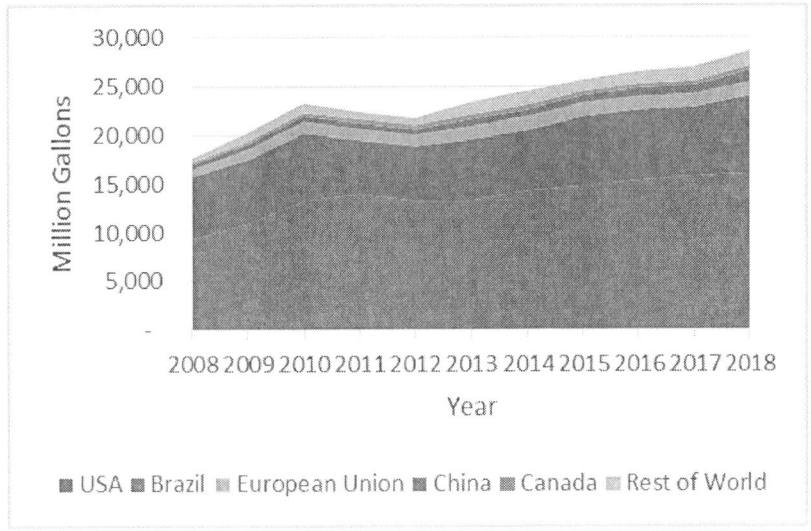

Figure 1. Worldwide production of ethanol (liquid biofuel) in the last decade (Based on data published in the homepage of Alternative Fuels Data Center and The Statistics Portal).

Fermentative processes used to produce bioethanol mainly from feed sources like corn starch and sugarcane juice (called first generation, where the conversion of glucose is commonly done by *Saccharomyces cerevisiae*) have several limitations in terms of land use, feed availability, and production costs (Dos Santos et al., 2016; Niphadkar et al., 2017; Passoth, 2017). Therefore, in a search for alternative technological processes, use of low-cost raw materials such as agricultural wastes and research on culture development and lignocellulosic fiber fermentation technology (Brooks et al., 1979), which had originally been proposed in the late 1970s, reemerged during the 21st-century, leading to second generation ethanol (2G) production.

The lignocellulosic biomass as a potential raw material is inexpensive, renewable, and abundant, and serves as a unique natural resource for large scale and cost-effective bio-energy generation (Anwar et al., 2014). Despite being a promising technology, the conversion of the sugars in the lignocellulosic biomass to ethanol includes great challenges due to the nature of the type of feedstock. Lignocellulose is a recalcitrant material and is the main constituent of the complex plant cell wall. The chemical composition varies according to the genetic and environmental characteristics (Balat, 2011), but it is estimated that the composition of the primary components ranges from 40–50% cellulose, 20–40% hemicelluloses, 20–30% lignin, and 5–8% extractives (Mosier et al., 2005; Limayem and Ricke, 2012; Anwar et al., 2014; Dos Santos et al., 2016). In the biochemical processes for the release of sugars capable of being fermented by microorganisms, the treatment of lignocellulosic biomass involves several steps (Figure 2) to remove lignin and break down glycosidic bonds to obtain monosaccharides as building blocks for ethanol production.

Although the process of producing ethanol from lignocellulosic material is promising, there are technical and economic disadvantages to its application for large-scale production. The formation of toxic compounds (organic acids, furan derivatives, and phenolic compounds) for the fermenting microorganisms during the sugar release process not only impacts the productivity of the process, but also increases the operating costs by introducing methods to mitigate the harmful effects in hydrolysis and fermentation. Depending on the pretreatment and hydrolysis methods used for the degradation of lignocellulosic raw materials, the carbon source present in the hydrolysate will not only be different, but may also give rise to polymers such as cellodextrins, xylodextrin, and cellobiose, or monomers such as glucose, xylose, arabinose, and acetate (Li et al., 2019). Due to the nature of the lignocellulosic material composition during its bioconversion, it is important to submit it for a prior treatment to obtain the polymeric components. Physical, chemical, and biological processes or combinations thereof have been proposed for this purpose in order to reduce the crystallinity of cellulose with the preservation of hemicellulose sugars and to minimize the production of inhibitory compounds (Chandel et al., 2018).

Physical treatment by means of chipping, crushing, and grinding processes, reduces the crystallinity of cellulose. The size of the materials is generally 10–30 mm after chipping and 0.2-2 mm after grinding or crushing (Sun and Cheng, 2002). Pyrolysis carried out at more than 300°C (Canilha et al., 2012) is another process used in the rapid degradation of cellulose to H_2, CO, and residual carbon.

Chemical treatment by acid processing (hydrochloric acid, phosphoric acid, sulfuric acid, and sulfur dioxide), increases the solubility of polysaccharides at different acid concentrations (Gírio, 2010). Hydrolysis with dilute acid under less severe conditions resulted in high conversion yields of xylan to xylose (Canilha et al., 2012). Alkaline pretreatment (sodium, potassium, calcium, and ammonium hydroxide) allows removal of lignin and several substituted uronic acids in the hemicellulose, by using low temperatures and pressures but the pretreatment time is hours or days (Balat, 2011). Treatment with organic solvent or organosolv (ethanol, ethylene glycol, methanol, and acetone) breaks down the lignin-lignin and lignin-carbohydrate bonds and promotes an increase in the surface area and volume, allowing enzyme access (Cardona et al., 2010).

Physicochemical treatment by vapor explosion or thermochemical method produces high yields of soluble hemicellulose (produces mainly oligosaccharides) with low solubility lignin (Canilha et al., 2012). For sugarcane bagasse, steam explosion pretreatment yields superior results for 5C sugar recovery in liquid and 6C sugar recovery in fiber, and for inhibitors, primarily acetic acid, formic acid, and lignin-derived phenolics (Chandel et al., 2018). The ammonia fiber explosion method is an alkaline thermal treatment that exposes the lignocellulosic material to a high temperature and pressure, followed by a rapid release of pressure (Sun and

Cheng, 2002); this pretreatment was used successfully for corn stover (Chandel et al., 2018).

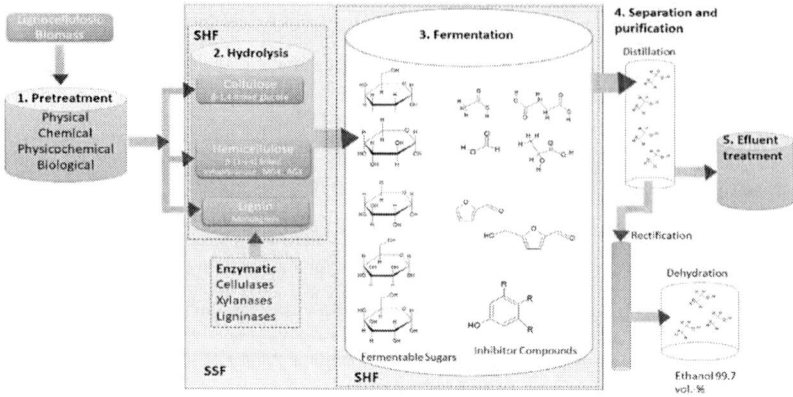

Figure 2. Biochemical pathways for bioconversion of lignocellulosic materials involves five steps: 1) pre-treatment, 2) hydrolysis of complex carbohydrates, 3) fermentation, 4) separation and purification, and 5) effluent treatment. MGX: O-acetyl-4-O-methylglucuronoxylans; AGX: arabino-4-O-methylglucuronoxylans. SHF: separate hydrolysis and fermentation SSF: simultaneous saccharification and fermentation.

Table 1. Enzymes used in the degradation of lignocellulosic biomass

Enzymes	Types
Cellulases	*Endoglucanases* (E.C.3.2.1.4)
	exoglucanases or cellobiohydrolases (EC 3.2.1.91)
	β-glucosidases (EC 3.2.1.21)
Xylanases	*endo^{-1},4-β-D-xylanases* (EC 3.2.1.8)
	1,4-β-D-xylosidases (EC 3.2.1.37)
	α-L-arabinofuranosidases (EC 3.2.1.55)
	α- glucuronidases (EC 3.2.1.1)
	acetyl xylan esterases (E.C. 3.1.1.72)
	p-coumric and ferulic acid esterases (E.C. 3.1.1.1)
Ligninases	*Laccases* (EC 1.10.3.2)
	lignin peroxidase (LIP) (EC 1.11.1.14)
	Manganese peroxidase (MnP) (EC 1.11.1.13)

Biological treatments use brown and soft bacteria, and white rot fungi to degrade lignin and solubilize hemicellulose (Balat, 2011). The enzymatic degradation of vegetable biomass is carried out by a complex mixture of enzymes (Table 1), among which the cellulases and hemicellulases stand out. In addition, ligninases are used to promote depolymerization of lignin (Martins et al., 2011).

The fermentation can be carried out as a batch process, batch feeding, or continuous. This process is carried out by microorganisms (yeasts, fungi, or bacteria) and the technological options suggest the use of a single microorganism or a microbial consortium (Cunha-Pereira et al., 2011; Gutiérrez-Rivera et al., 2012; Hickert et al., 2013). Selection of microorganisms for ethanol production depends on the desired level of fermentative process parameters such as ethanol yield (> 90–95% theoretical yield (equivalent 0.45–0.48 $g.g^{-1}$ sugar)) (Hahn-Hägerdal et al., 2007), tolerance to ethanol (> 40% g L^{-1}), ethanol productivity (1 g $L^{-1}.h^{-1}$), low-cost medium formulation, resistance to inhibitors and acid, and thermal stress conditions (Senthilkumar and Gunasekaran, 2005).

Compared to other types of microorganisms, yeasts especially *S. cerevisiae,* which is a GRAS (Generally Recognized as Safe) microbe, is employed in ethanol production due to its high ethanol production, high ethanol tolerance, and the ability to ferment a wide range of sugars (Azhar et al., 2017). Crabtree-positive yeasts like *S. cerevisiae* have the ability to transform 6C sugar (glucose) into 2C (ethanol) compounds efficiently, however, in the presence of 5C sugar (xylose) the catabolic (glycolysis) and anabolic (gluconeogenesis) metabolism present some variations (Kwak et al., 2018). Thus, the inability of this yeast to ferment 5C sugar has been a barrier for the implementation of existing technology. In the glucocentric approach (Dumon et al., 2012) for first-generation ethanol production, *S. cerevisiae* is inefficient to take advantage of the proportion of pentoses present in the lignocellulosic material.

An important factor for the viability of the biomass transformation process is the need to convert all the sugars in the hydrolysate to ethanol, a process in which microorganisms metabolically transform sugars of five (5C) and six carbons (6C) to 2C ethanol. Compared to glucose, xylose was recognized as a Crabtree-negative carbon source (Li et al., 2019). Xylose metabolism has the advantage to prevent ethanol accumulation by overcoming the Crabtree effect in yeast. However, less stringent pathway regulation during xylose metabolism relaxes the metabolic flow, such that cell metabolism is redirected from strictly fermentative metabolism of glucose to respiratory metabolism of xylose, resulting in carbon loss in the form of CO_2 release (Li et al., 2019). For this reason, efforts to find natural robust microorganisms or yeast constructs with metabolic capacities that meet the demands of the fermentation processes of the cellulose fraction such as hemicellulose, have led to an increase in description of microorganisms with natural ability to ferment pentoses, mechanisms associated with resistance under stress conditions, as well as advances in tools for the development of genetically modified microorganisms. In the last year, natural xylose-fermenting yeasts have been described (Table 2) with the potential to transform sugars into ethanol from different sources of residual biomass.

2. Xylose Metabolism by Yeasts

For the fermentation process, yeasts are the preferred microorganism for use as biocatalysts in industrial processes because they exhibit basic characteristics such as, reproduction by budding or fission, robustness, and resistance against low pH, tolerance for high osmotic pressure and phage infection (Kwak et al., 2018).

2.1. Xylose Transporters

Description of several yeast species which naturally ferment xylose aroused great interest in these phenotypes for the transformation of the hemicellulosic portion of the residual biomass. In the catabolism of D-xylose, pentose uptake rate is important for ethanol production, and affinity of the transporter for the sugars plays a key role. Active and passive transport processes contribute to the uptake of this sugar and the sugar is transported by a low-affinity facilitated xylose transporter or by the high-affinity xylose/H+ symporter. With very few exceptions, symporters are expressed at low-sugar concentrations while the low-affinity facilitative transporters are expressed when the sugar is abundant (Leandro, Fonseca, and Gonçalves, 2009).

The hexose transporter (HXT) family of homologous genes is described widely, and they encode the hexose transporters expressed by *S. cerevisiae*. This family includes 18 transporters (HXT1–HXT17, and GAL2) and two putative sensors (SNF3 and RGT2) (Diderich et al., 1999). Several of these carrier proteins are involved in the uptake of xylose for entry into the cell by facilitated diffusion, mediated by hexose transporters (Hamacher et al., 2002; Farwick et al., 2014). A study of both hxt-null mutant of *S. cerevisiae* and a wild-type strain genetically modified to express the *XYL1* gene coding for xylose reductase (XR) from *Sc. stipitis*, showed that wild-type *S. cerevisiae* exhibits better xylose consumption rate than hxt-null mutant strain (without eighteen hexose transporters) (Kim et al., 2017).

Although, *S. cerevisiae* does not have the metabolic machinery for the fermentation of xylose, its transporters have been a focus of interest in the transport of this pentose, especially the high-affinity transporters (HXT2, HXT4, HXT7, and Gal2) that are active in the transport of xylose (Knoshaug et al., 2015, Miskovic et al., 2017). Mutations in asparagine 376/370 or threonine 219/213 of transporters Gal2 and Hxt7, respectively, showed that changes in a confined region probably create a steric hindrance for hexoses that result in an increased affinity for D-xylose and a decreased or total loss of affinity for D-glucose (Farwick et al., 2014).

The search for pentose transporters in naturally fermenting yeasts of D-xylose and L-arabinose has also been used as a strategy in metabolic engineering. However, studies showed that some transporters, when expressed in *S. cerevisiae*, do not behave as they do in their native organism. The gene *KmAxt1p* from *Kluyveromyces marxianus* (*K. marxianus*) expressed in *S. cerevisiae* showed low-affinity transport kinetics for L-arabinose (Km=263 mM and Vmax=57 nM/mg/min) and the transport is inhibited by glucose, galactose, and xylose (Knoshaug et al., 2015).

Table 2. Non-Conventional yeast strains with potential to be used in bioethanol (2G) production

Yeast Strain	Type of Strain	Feedstock (detoxification)	Sugars	System and fermentation conditions	Sugar concentration (g.L^{-1})	Ethanol Yield (g.g^{-1})	Reference
Scheffersomyces stipitis NRRL-Y-7124	WT[a]	Coffee silverskin[b] (Neutralization)	Glucose, arabinose, galactose and mannose	Batch[e] 30°C, 200 rpm, 48 h	50	0.26	Mussatto, et al. (2012)
	WT	Spent coffee grounds[b] (Neutralization)	Glucose, Xylose arabinose, galactose and mannose	Batch[e] 30°C, 200 rpm, 48 h	20	0.11	
Spathaspora passalidarum UFMG-HMD-1.1	WT	Sugarcane bagasse[b] (Overliming)	Glucose, xylose, and rabinose	Batch[e] 30°C, 200 rpm, 96 h	61	0.20	Cadete et al. (2012)
Scheffersomyces lignosus CBS 4705	WT				61	0.16	
Spathaspora suhii UFMG-XMD-16.2	WT				61	0.23	
Spathaspora roraimanensis UFMG-XMD-23.2	WT				61	0.22	
Scheffersomyces stipitis UFMG-XMD-15.2	WT			Batch[e] 30°C, 200 rpm, 72 h	61	0.34	
Spathaspora hagerdaliae UFMG-CM-Y303	WT	Sugarcane bagasse[c] (Non-detoxified)	Glucose and xylose	Batch[e] 28°C, 180 rpm, 48 h	18.5	0.29	Rech et al. (2019)
				Batch[f] 28 °C, 180 rpm, 48 h	18.5	0.24	

Yeast Strain	Type of Strain	Feedstock (detoxification)	Sugars	System and fermentation conditions	Sugar concentration (g.L^{-1})	Ethanol Yield (g.g^{-1})	Reference
Kluyveromyces marxianus CE025	WT	Cashew apple bagasse[b] (Neutralization)	Glucose, xylose, arabinose	Batch[e] 30°C, 200 rpm, 48 h	57	0.34	Rocha et al. (2011)
Scheffersomyces shehatae NCIM 3501	WT	Sugarcane Bagasse[b] (Neutralization)	Glucose, xylose, arabinose	Batch[e] 30°C, 150 rpm, 24 h	20	0.22	Chandel et al. (2007)
		Sugarcane Bagasse[b] (Activated charcoal)			20	0.42	
Scheffersomyces stipitis CBS6054	WT	Giant reed[b] (Neutralization)	Xylose, glucose, arabinose	Batch[e] 30°C, 150 rpm, 96 h	25	0.33	Scordia et al. (2012)
Spathaspora passalidarum UFMG-HMD-14.1	WT	Sugarcane bagasse[d] (Non-detoxified)	Cellobiose, glucose, xylose, and arabinose	Batch[e] 30°C, 120 rpm, 72 h	100	0.32	De Souza et al. (2018)
Scheffersomyces parashehatae UFMG-CM-Y507	WT	Sugarcane bagasse[b] (Overliming)	Glucose, xylose arabinose	Batch[e] 30°C, 200 rpm, 72 h	61	0.25	Cadete et al. (2017)
Scheffersomyces illinoinensis UFMG-CM-Y513	WT			Batch[e] 30°C, 200 rpm, 48 h	61	0.23	
Spathaspora arborariae UFMG-CM-Y352	WT			Batch[e] 30°C, 200 rpm, 96 h	61	0.16	

[a]Wild-type; [b]Hydrolysate from acid treatment; [c]Hydrolysate from steam explosion and enzymatic hydrolysis; [d]Hydrolysate from alkaline peroxide and enzymatic hydrolysis; [e]SHF (separate hydrolysis and fermentation); [f]SSF (simultaneous saccharification and fermentation).

The presence of other sugars strongly affects pentose transporters. Three genes coding for transporters of D-xylose *XUT1*, *HXT2.6*, and *QUP2* of *Sc. stipitis* were identified and cloned separately in hxt-null strains of *S. cerevisiae* that demonstrated a 40-50% decrease in the consumption of D-xylose when D-glucose was present in the fermentation medium (Belisa et al., 2015). On the other hand, the gene *PgAxt1* from *Meyerozyma guilliermondii* cloned in *S. cerevisiae* (that showed same kinetic characteristic for L-arabinose of the wild-type strain *M. guilliermondii*) maintains the transport of xylose to almost 90% of the normal levels when competing with arabinose (Knoshaug et al., 2015).

Active transport systems are also under consideration for engineering efficient microorganisms for transport of D-xylose. Chromosomal integration of multiple copies of the *araE* gene (an arabinose/H+ symporter) from *Bacillus subtilis* and a single copy of the *XYL1* gene (encoding xylose reductase (XR)) from *Sc. stipitis* in *S. cerevisiae* resulted in 18.7 g/L xylose consumption with a rate of 0.39 g/L-h in a batch culture (Kim et al., 2017). Nevertheless, the active transporters also are competitively inhibited due to sugar affinity. The proton symporter XylE from *E. coli*, that is exceptionally selective for xylose like most symporters, has a high affinity for xylose (Kd Xyl = 0.35 mM) and also exhibits affinity for glucose (Kd Glc = 0.77 mM) (Farwick et al., 2014). The positive influence of XR on the uptake of xylose was also shown by Kim et al. (2017) and additional expression of *XYL1* in the *S. cerevisiae* mutant strain (containing *araE* and *XYL1* gene) showed a 1.4-fold increase in all parameters for xylose consumption.

Functional activity of the transporters in central catabolism of carbohydrates was proposed based on the metabolic profiles observed when different transporters of *Sc. stipitis* were expressed independently in *S. cerevisiae*. Mutants with the *XUT1* gene produced more ethanol than mutants with the *HXT2.6* and *QUP2* gene (from *Sc. stipitis*), where the decrease in ethanol concentration was mainly due to the formation of byproducts such as xylitol and glycerol (Belisa et al., 2015). From the biology of systems, the implementation of kinetic modeling programs has also confirmed that the modification of enzymatic activity improves the metabolic fluxes for the transport of xylose. Deletion of the gene for hexokinase (*HXK2*) resulted in an improvement of the rate of uptake of xylose by hexose transporters present in *S. cerevisiae* (Miskovic et al., 2017).

2.2. Reductive/Oxidative XR-XDH Pathway

After the transport of D-xylose from the extracellular medium into the cell, the main metabolic pathway is the pentose phosphate pathway. In this pathway, xylulose-5-phosphate (xylulose-5P) acts as an intermediate metabolite in the non-oxidative phase. Transketolases and transaldolases participate in the isomerization of xylulose-5P to its products which are channeled into the central carbon metabolism pathways such as glycolysis, gluconeogenesis, the Krebs cycle, and other pathways that lead to maintenance of cellular activity through the generation of intermediates to provide energy to the microorganism (Kwak and Jin, 2017)

Conversion of D-xylose to xylulose-5P is catalyzed by two oxidoreductases that play an important role in redox balance of cofactors such as NADH, NADPH, NAD^+, and $NADP^+$. XR, (EC 1.1.1.30) is the first enzyme in this pathway that participates in catalyzing the conversion of D-xylose to xylitol and controls the rate of xylose utilization. Many of them are NADPH-dependent however, NADH also serves as a coenzyme for some XR, but the biological advantage of XRs that utilize both NADPH and NADH as coenzymes is not yet understood (Hossain et al., 2018). The yeast XR exhibits differences in kinetic parameters, with a wide range of values for its kinetic constants i.e., K_{cat} from 860 min^{-1} (from *S. cerevisiae*) to 3600 min^{-1} (from *N. crassa*), Km for xylose from 13.6 mM (from *S. cerevisiae*) to 72 mM (from *C. tenius*) and Km for NADPH from 1.8 µM (from *N. crassa*) to 56 µM (from *C. intermedia*) (Hossain et al., 2018).

The next step in the pathway is the oxidation reaction of xylitol to xylulose that is catalyzed by the enzyme NAD^+-linked xylitol dehydrogenase (XDH, EC 1.1.1.9). During anaerobic fermentation of xylose in yeasts using the pathway XR/XDH, an accumulation of NADH occurs due to a preference of XR and XDH toward NADPH and NAD$^+$, respectively. Since the accumulated NADH cannot be oxidized without oxygen, its metabolism is blocked (Hossain et al., 2018).

The redox couples NAD$^+$/NADH and NADP$^+$/NADPH are essential for maintenance of cellular redox homeostasis (equivalent to pyruvate NADH/NADPH 1.6/0.6) (Li et al., 2019) and for modulating numerous biological events, including cellular metabolism. An imbalance in these factors affects multiple catabolic and anabolic pathways leading to a significant change in the metabolic profile of a strain (Kwolek-Mirek, Maslanka, and Molon, 2019). Maintenance of redox homeostasis has a significant impact not only on cell metabolism but also on many other processes including stress responses, intracellular signaling, protection of protein thiols, redox reactions as well as removal of xenobiotics (Ayer et al., 2014). Maintenance of NADPH is crucial for the supply of cellular GSH (glutathione reduced form) and for optimal redox balance in the cells. The cellular demand of NADPH increases especially in proliferating cells where DNA replication forces the synthesis of nucleotides independent of the synthesis of proteins and lipids (Kwolek-Mirek, Maslanka and Molon, 2019).

To overcome the imbalance in cofactors, several strategies were attempted to increase the xylose utilization efficiency. Mutagenesis studies were carried out to discriminate the binding of NAD(H) and NADP(H). Mutation of XR form *Sc. stipitis* using site-directed mutagenesis to introduce single or double mutations at specific positions in the primary sequence of the enzyme in the Lys270, Asn272 and Arg276 residues showed the involvement of these residues in coenzyme selectivity (Hossain et al., 2018). Also, random mutagenesis showed the importance of amino acids at positions between 260 and 280 in discriminating the binding of NAD(H) and NADP(H) (Hossain et al., 2018).

2.3. Direct Isomerization of Xylose to Xylulose by Xylose Isomerase (XI) Pathway

In this pathway, the XI, (EC. 5.3.1.5) converts xylose to xylulose without involvement of coenzyme (equivalent to pyruvate NADH / NADPH 1/0) (Li et al., 2019). It carries out bioconversions under anaerobic and oxygen limiting conditions to maintain the redox balance and therefore has received a lot of attention. However, the inherent thermodynamic equilibrium between the substrate and the product leads to low conversion yields and creates difficulties for the separation and purification of the byproducts (Liu et al., 2019). Due to difficulties in post-translational modifications and protein mis-folding (Hossain et al., 2018) only small copy number of XI are expressed in yeasts.

To overcome the thermodynamic limitations in isomerization reactions, increasing the efficiency of endogenous carbon metabolism by increasing transcription of genes from the pentose phosphate pathway or molecular chaperones, which stimulate metal-dependent XI activity, or the biogenesis of the Fe-S group, has been a focus of yeast metabolic engineering (Li et al., 2019). Expression of the XI on the cell surface has also been proposed as a useful strategy for ethanol production from xylose because xylulose could easily be transported into the cell (Ota et al., 2013).

A comparison of the XI pathways in *Piromyces* and XR / XDH from *Sc. stipitis* expressed independently in recombinant strains of *S. cerevisiae* showed that xylose utilization via the XI pathway is efficiently directed towards the production of ethanol (ethanol yield 0.43 g.g^{-1}) and results in low levels of by-products. However, the rate of consumption of xylose (0.05 g xylose g cells^{-1}h^{-1}) is lower in the strain with the XR/XDH pathway (0.13 g xylose g cells^{-1}h^{-1}) where the production of ethanol was lower (yield of ethanol 0.33 g.g^{-1}) due to the formation of by-products such as xylitol and glycerol (Karhumaa et al., 2007). The XI (from *Clostridium phytofermentans*) and XR/XDH (from *Sc. stipitis*) were also expressed and an evaluation of the activity in synthetic media and in the presence of oxygen, showed superiority of the strains expressing XI for ethanol production, while the strains with the XR/XDH pathway accumulated higher xylitol levels, however, the rates of consumption of xylose for the two strains (containing XI or XR/XDH) were similar (Cunha et al., 2019). These results show production of higher levels of by-products through the XR/XDH than the XI pathway. When Cunha et al. (2019) expressed the XI and XR/XDH pathways together, the strains of *S. cerevisiae* showed a lower ethanol yield than the strain containing only the XI, due to the formation of by-products possibly derived from the XR/XDH pathway.

3. NATURAL XYLOSE-FERMENTING YEASTS

Xylose-fermenting yeasts are found in rotting wood, tree barks, leaves, and some are associated with insects (Figure 3) (Cadete, Lopes, and Rosa, 2017). Bodin and Adzet discovered Pachysolen tannophilus (*Pa. tannophilus*) from a wood extract in 1957 which was the first yeast shown to convert D-xylose into ethanol, gaining scientific notoriety since no yeast was able to ferment aldopentoses (Kurtzman, 1983) until then.

Figure 3. (a) Beetles in dead wood and (b) rotting wood in Amazonian Rainforest.

Fermentation of D-xylose by *Pa. tannophilus* when tested under different aeration conditions showed a maximum yield of 52% of ethanol from xylose under oxygen-limiting conditions. No ethanol was produced under anaerobic conditions (Schneider et al. 1981) and similar results by

Slininger et al. (1982) showed lack of ethanol production in the absence of oxygen when a lower quantity of cell inoculum was used. When the inoculum amount increased, ethanol was produced despite lack of cell multiplication (Kurtzman, 1983).

Although alcoholic fermentation is traditionally anaerobic, many yeasts have the ability to ferment in the presence of oxygen (Correia et al. 2018). The oxidoreductase enzymes, XR and XDH, have different affinities for the cofactors involved in the metabolism of xylose. *Pa. tannophilus* and many of xylose-fermenting yeasts have, XR dependent on NADPH or NADH, with a preference for NADPH and XDH dependent on NAD^+ or $NADP^+$ but with a higher affinity for NAD^+ (Ditzelmüller et al. 1984; Ditzelmüller et al. 1985). Therefore, the conversion of xylose to xylulose implies the production of $NADP^+$ and NADH. However, the regeneration of NADPH and NAD^+ is necessary to maintain the redox balance. NADPH is generated by directing part of the fructose-6-phosphate produced in the oxidative step to the pentose phosphate pathway. Under aerobic conditions, NADH can be re-oxidized via respiratory chain with molecular oxygen. Under anaerobic condition, an electron acceptor is required to re-oxidize NADH or the cell is unable to maintain the redox balance resulting in xylitol accumulation in the cell (Ditzelmüller et al., 1984; Verduyn et al., 1985).

Many yeasts of the Saccharomycotina subphylum are xylose-fermenting, but the highest fermentation rates were seen in species of the genus *Scheffersomyces* (Urbina and Blackwell, 2012). Based on multilocus analysis with the small subunit (SSU) and large subunit (LSU) rRNA markers, RNA polymerase II subunit 1 (RPB1), and internal transcribed spacer region (ITS), this genus as proposed by Kurtzman and Suzuki (2010) contains species previously harbored by the genus *Pichia* (*P. stipitis, P. segobiensis,* and *P. spartinae*) and *Candida* species (Lopes et al., 2018). *Scheffersomyces* species inhabit environments rich in D-xylose and are generally found in decaying wood, insect larvae, and wood beetles (Cadete, Lopes and Rosa, 2017). Ethanol producing species of the genus are, *Sc. stipitis, Sc. shehatae, Sc. lignosus, Sc. insect, Sc. illinoinensis, Sc. cryptocercus,* and *Sc. virginianus* (Ren et al., 2014).

For several years, *Sc. stipitis* was the highest ethanol-producing yeast among xylose-fermenting yeasts. For this reason, it is used worldwide for the production of ethanol from pentoses (du Preez and van der Walt, 1983; du Preez, Van Driessel, and Prior, 1989; Agbogbo and Coward-Kelly, 2008). In fed-batch cultures, *Sc. stipitis* was able to produce up to 47 g L^{-1} of ethanol from xylose at 30°C resulting in ethanol yields in the range of 0.35-0.44 g.g $^{-1}$ (du Preez et al. 1989; Hahn-Hägerdal and Pamment, 2004). du Preez et al. (1986) investigated the conversion of D-xylose into ethanol by *Sc. shehatae* and *Sc. stipitis* species at different temperatures and pH ranges. For both yeasts the optimal pH range was 4.0 to 5.5, the optimal fermentation temperature was 30°C, and maximum ethanol was produced with 50 g L^{-1} of xylose concentration. It was shown that *Sc. stipitis* and *Sc. shehatae* used 100% of xylose present in the medium below 36°C. At 36°C, *Sc. shehatae* consumed only 27% of the sugar while *Sc. stipitis* consumed 87% of the xylose present in the medium being more thermotolerant than *Sc. shehatae*. Among the yeast species *Pa. tannophilus*, *K. marxianus*, *Sc. shehatae*, and *Sc. stipitis*, the last one showed promise in the production of 2G ethanol (Delgenes, Molleta, and Navarro, 1986). *Sc. stipitis* is able to ferment an extensive range of sugars, including cellobiose, to produce high ethanol yields. However, limiting oxygen levels is essential for the yeast to maintain viability and balance redox without producing detectable xylitol (Grootjen, Van der Lans, and Luyben, 1990; Agbogbo et al. 2006).

The *XYL1* and *XYL2* genes of xylose metabolism in *Sc. stipitis* are not expressed when glucose is present in the medium, as in lignocellulosic hydrolysates (Jeffries et al., 2007). Similar to most xylose-fermenting yeasts, *Sc. stipitis* primarily consumes glucose instead of xylose since the presence of glucose inhibits xylose uptake (Slininger, Bolen, and Kurtzman, 1987). *Sc. stipitis* and *Sc. shehatae* have two transport systems, facilitated diffusion transporters (low-affinity) and proton symporters (high-affinity) for xylose uptake (Kilian and Van Uden, 1988). *Sc. stipitis* has specific genes for high-affinity xylose transporter, *XUT1* to *XUT7*, as well as genes expressing hexose transporters with lower transport capacity, *RGT2*, *SUT1*, *SUT2*, *SUT3*, and, *SUT4* (Weierstall et al. 1999; Young et al. 2011; Moon et al. 2013). However, glucose competes with xylose for the low-affinity system and inhibits xylose transport by the high-affinity system non-competitively (Gonçalves et al. 2004; Sharma et al. 2018). Using different ratios of glucose to xylose (Agbogbo et al. 2006) or just one sugar in the culture medium, observed the highest concentration of cells (7.95 g L^{-1}) with 100% glucose in the medium. At a ratio of 75% glucose/25% xylose, glucose was consumed by 72 h of fermentation, before xylose consumption began, reaching a final ethanol concentration of 22.3 g L^{-1}. With a ratio of 25% glucose/75% xylose, the ethanol concentration was 24.4 g L^{-1} and all the glucose was consumed in 24 h, showing that the glucose concentration should be less than 2% (w/v) before significant consumption of xylose is detected when combination of sugars are used (Panchal et al. 1988).

Recently, *Spathaspora passalidarum* and *Spathaspora arborariae* species were isolated and identified (Nguyen et al., 2006; Cadete et al., 2009), with promising results for the production of ethanol from D-xylose, especially *Sp. passalidarum* (Nguyen et al. 2006; Cadete and Rosa, 2018). The genus *Spathaspora* described by Nguyen et al. (2006) was from a single strain, *Sp. passalidarum*, initially isolated from the gut of the *Odontotaenius disjunctus* beetle belonging to the Passalidae family, in Louisiana — USA (Nguyen et al. 2006). Phylogenetic analysis of the complete genome of *Sp. passalidarum* showed that it belongs to the CUG clade of yeasts that decode serine instead of leucine, in which *Sc. stipitis* is included (Wohlbach et al. 2011). Cadete et al. (2009) isolated a second species of the clade, *Sp. arborariae,* from decaying wood collected in Atlantic Forest and Cerrado ecosystems in Brazil. Nowadays, there are many described species of the genus: *Sp. brasiliensis, Sp. roraimensis, Sp. xylofermentans, Sp. piracicabensis, Sp. hagerdaliae, Sp. suhii, Sp. girioi, Sp. gorwiae, Sp. Allomyrinae,* and *Sp. boniae* (Cadete et al. 2009; Cadete et al. 2012; Lopes et al. 2016; Wang et al. 2016; Morais et al. 2017; Varize et al. 2018).

Sp. passalidarum ferments xylose efficiently and achieves high yields of ethanol (up to 0.48 $g.g^{-1}$), close to the theoretical maximum value (0.51 $g.g^{-1}$), in addition to rapid consumption of sugars (Hou, 2012; et al., 2006; Cadete et al., 2016). Other species of the clade, such as *Sp. brasiliensis, Sp.*

Roraimensis, and *Sp. xylofermentans* being xylitol producers have XR with a stringent requirement for NADPH; while the XR of ethanol-producing *Sp. passalidarum* and *Sp. arborariae* has an affinity for both NADH and NADPH cofactors. However, XR from *Sp. arborariae* was more dependent on NADPH. Even under anaerobic conditions, *Sp. passalidarum* has higher ability to ferment xylose due to balance in cofactors due to the activity of XR and XDH (Cadete et al. 2016; Cadete and Rosa, 2018). *Sp. passalidarum* has two genes, *XYL1.1* and *XYL1.2*, which encode XR with a preference for NADPH and NADH, respectively. Thus, the cofactors are balanced actively and metabolic reactions lead to the pentose phosphate pathway (PPP) and glycolysis pathway for ethanol production (Wohlbach et al. 2011; Cadete et al. 2016). Currently, *Sp. passalidarum* is the ideal yeast to convert D-xylose into ethanol (Cadet and Rosa, 2018).

Comparison of *Sp. passalidarum* and *Sc. stipitis* species under different fermentation conditions showed a high yield of ethanol production by *Sp. passalidarum* under anaerobic conditions, with xylose being consumed after glucose (Hou, 2012). Starting with 32 g L^{-1} xylose the consumption of sugar was 99%, more than 90% of the theoretical maximum value of ethanol and a concentration of 14 g L^{-1} was reached. The enzymatic activity of XR was preferentially dependent on NADH, resulting in a balance of essential cofactors in this pentose metabolism pathway. The yeast *Sc. stipitis* was not able to use xylose under anaerobic conditions. Cadete et al. (2016) used species of the genus *Spathaspora* to study xylose metabolism under oxygen-limiting conditions. Xylose fermentation was characterized using the species, *Sp. passalidarum*, *Sp. arborariae*, *Sp. brasiliensis*, *Sp. roraimanensis*, *Sp. suhii*, and *Sp. xylofermentans*. The results showed that strains of *Sp. passalidarum* were the best xylose-fermenting yeast. From 40-50 g L^{-1} of D-xylose, pentose consumption was 99.4% with the production of 20.5 g L^{-1} of ethanol in 24 h. The XR cofactor preference of this species was checked using the NADH as a cofactor. In *Sp. passalidarum* strains, a balance in cofactors are associated with a high yield of ethanol production during conversion of xylose. This species also consumes glucose and xylose simultaneously under aerobic conditions, with ethanol yields above 0.42 $g.g^{-1}$ (Hou, 2012; Su et al.

2015). In detoxified sugarcane bagasse hydrolysate, *Sp. passalidarum* consumed all the glucose content and more than 90% of the xylose with 3-fold higher glucose in the medium than pentose. This yeast had a high yield of ethanol (0.46 g.g^{-1}) and volumetric productivity (0.81 g L^{-1} h^{-1}), while *Sc. stipitis* yielded 0.32 g.g^{-1} ethanol and volumetric productivity was 0.36 g L^{-1} h^{-1}. There is no information on the xylose transport system of *Sp. passalidarum* and no explanation as to why this species uses hexose and pentose simultaneously. The researchers suggest the presence of a low affinity facilitated diffusion transporter and a high-affinity xylose H$^+$ symporter for uptake of sugars, as previously described (Leandro, Fonseca, and Gonçalves, 2009; Rodrussamee et al., 2018).

Recently, a strain of *Sp. passalidarum* CMUWF1-2 was isolated from soil samples in the Mae Taeng District, Chiang Mai Province, Thailand (Rodrussamee et al., 2018). This yeast was able to grow at 42°C in medium containing different sugars (mannose, galactose, xylose, glucose, and arabinose) and produce ethanol at three temperatures (30, 37, and 40°C) from the same sugars, except arabinose. The highest ethanol yield was reached with medium containing only xylose as substrate 0.43, 0.40, and 0.20 g.g^{-1} at 30, 37, and 40°C, respectively. In media, with mixtures of 20 g L^{-1} glucose / 20 g L^{-1} xylose, the ethanol yields were 0.42, 0.39, and 0.33 g.g^{-1} at 30, 37, and 40°C, respectively. The yeast also showed resistance to 0.2% of 2-deoxyglucose (2-DOG) present in media with the sugars mentioned above (Rodrussamee et al., 2018). The differences between this strain and that described by Nguyen, is that the latter does not assimilate arabinose, is capable of producing ethanol at 25–32°C, and xylose consumption was inhibited with 2-DOG at 0.1% (Hou, 2012; Rodrussamee et al., 2018).

Other *Spathaspora* species that were isolated from rotting wood and able to ferment xylose and produce ethanol was also described. Lopes et al. (2016) described three *Spathaspora* species, two of which are ethanol producers: *Sp. gorwiae* and *Sp. hagerdaliae*. Fermentation assays using D-xylose as the only carbon source, the strain *Sp. gorwiae* UFMG-CM-Y312 consumed 76.9% of the sugar achieving an ethanol yield of 0.10 g.g^{-1} in 60 h of fermentation. When the strain was cultured in both xylose and glucose, it consumed 86.7% of glucose and 20.4% of xylose in 24 h (0.24 g.g^{-1}). Two strains of *Sp. hagerdaliae*, UFMG-CM-Y304, and UFMG-CM-Y424 used 99.3 and 94.3% D-xylose, respectively and resulting in ethanol yields of 0.28 and 0.25 g.g^{-1}, respectively. These strains consumed all the glucose and 99% of xylose, with ethanol yield of 0.37 and 0.36 g.g^{-1}, respectively. The XR of *Sp. hagerdaliae* was strictly dependent on NADPH, while XR of *Sp. gorwiae* used both cofactors (Lopes et al. 2016). The strains of *Sp. piracicabensis* UFMG-CM-Y5867^T and UFMG-CM-Y6112 produced 3.37 and 3.48 g L^{-1} of ethanol in 24 h from 2% of xylose. However, *Sp. passalidarum, Sp. arborariae,* and *Sc. stipitis* are able to produce high titers of ethanol from D-xylose, with ethanol yields up to 0.37 g.g^{-1}, in similar experimental conditions (Cadete et al. 2009; Cadete et al. 2012).

In addition to the species of the genera *Spathaspora* and *Scheffersomyces*, there are other genera that convert xylose into ethanol. *K. marxianus*, one of the first yeast species described to produce ethanol from D-xylose, was grown in a medium containing 20 g L^{-1} of xylose and achieved 5.6 g L^{-1} of ethanol concentration (yield 0.28 g.g^{-1}) in 48 h (Margaritis and Bajpai, 1982). From Japanese cedar hydrolysate, the thermotolerant strain of species *K. marxianus* DMB1 produced 19.78 (yield: 0.45 g.g^{-1}), 25.98 (0.60 g.g^{-1}), and 13.40 g L^{-1} (0.35 g.g^{-1}) of ethanol at 30, 42, and 45°C, respectively. The same strain produced 23.74 (0.35 g.g^{-1}), 8.92 (0.49 g.g^{-1}), and 2.47 g L^{-1} (0.34 g.g^{-1}) of ethanol at 30, 42, and 45°C, respectively, from eucalyptus hydrolysate (Goshima et al. 2013). Several species of the *Sugiyamaella* clade were associated with rotting plant materials, soil, and gut of insects (Urbina et al. 2012). Sena et al. (2017) showed three strains of *Su. xylolytica*, one strain of *Su. valenteae*, and a single strain of *Su. bahiana, Su. bonitensis, Su. boreocaroliniensis, Su. lignohabitans,* and *Su. xylanicola* were able to produce ethanol from D-xylose, with a maximum yield of 0.177 g.g^{-1}. *Su. bonitensis* UFMG-CM-Y608 produced the highest titers of ethanol (4.51 g L^{-1}), and showed the highest ethanol yield of 0.177 g.g^{-1}, corresponding to 37.18% of fermentation efficiency, at 72 h. *Su. xylolytica* UFMG-CM-Y2135 showed the highest ethanol productivity (0.104 g L^{-1} h^{-1}) and a yield of 0.117 g.g^{-1}

at 24 h. Ethanol production from D-xylose by *Su. boreocaroliniensis* and *Su. xylanicola* was 1.77 and 1.83 g L^{-1} of ethanol, respectively. A single strain of *Wickerhamomyces* sp. (UFF-CE-3.1.2), isolated from rotting wood, reached 11.60 g L^{-1} of ethanol in 72 h of fermentation in non-detoxified and diluted 1:3 (v/v) sugarcane bagasse hydrolysate, containing 2.57 g L^{-1} of acetic acid (Bazoti et al., 2017).

In recent years, isolation of non-conventional yeasts capable of xylose uptake to produce ethanol as the main fermentation metabolite has increased. These yeasts can be acclimatized to resist inherent difficulties in the alcoholic industrial fermentation, such as microbial inhibitors of the hydrolysate and ethanol concentration (Wang et al. 2018). In addition, these yeasts have genes that express essential proteins for directing the metabolism of D-xylose into ethanol, such as *XYL1*, *XYL2*, and *XYL3* (Cadete and Rosa, 2018). Several reports show the use of metabolic engineering tools to insert yeast genes from *Pa. tannophilus*, *Sc. stipitis* and *Sp. passalidarum* in *S. cerevisiae* strains (Kim et al. 2012; Cadete et al. 2016; Cunha et al. 2019). Non-conventional yeasts from several habitats and their unique traits are a source of biocatalysts for direct application in industrial processes.

4. XYLANOLYTIC ENZYME PRODUCERS

Xylan is a linear chain of D-xylopyranosyl residues linked by β-1,4-glycosidic bonds and is the most abundant compound present in hemicellulose. The xylanolytic enzymes include: endo-β-xylanase and β-xylosidase, acetyl xylan esterase, feruloyl esterase, arabinofuranosidase, glucuronyl esterase, glucuronidase, glucanase, mannanase, xyloglucanhydrolase, and monooxygenases (Kirikyali and Connerton, 2015; Prajapati et al., 2018). The enzymatic system that degrades hemicellulose is more complex than cellulose because of a variety of polysaccharide structures present in the hemicellulosic fraction (Sharma et al., 2019). The enzymes endo-1,4-β-xylanase and β-xylosidase are the major types of hydrolytic enzymes responsible for degrading xylan.

Basically, 1,4-β-xylanase catalyzes the hydrolysis of xylan releasing xylooligosaccharides, while β-xylosidase acts at the reducing terminal and releases D-xylose (Kumar and Shukla, 2016; Moreira, 2016). The development of research on microorganisms that degrade xylan and its enzymatic systems has become increasingly relevant in both economic and ecological terms (Singh et al., 2019). Hydrolytic enzymes are produced by saprophytic microorganisms that grow in decaying materials, and soil and water samples serve as sources for protein and enzyme isolation (Juturu and Wu, 2013). Although filamentous fungi are known to be great xylanolytic enzyme producers, many yeast species were described for their enzyme production (Cadete et al., 2009, Sena et al., 2017).

Aureobasidium pullulans (*A. pullulans*) is an anamorphic and ubiquitous yeast-like, which is found in different habitats such as soil, decaying materials, rock surfaces, and aquatic environments (Prasongsuk et al., 2018). It is recognized for its diverse biotechnological potential due to the production of several extracellular enzymes, including xylanases, β-glucosidases, cellulases, amylases, proteases, and mannoses (Bozoudi and Tsaltas, 2018). Color variant strains of *A. pullulans* are prominent endoxylanase producers (EC 3.2.1.8) and exhibit brilliant pigments with red, yellow, orange, or purple coloring, different from the black or off-white coloring of typical strains. Typically, pigmented and color-variant strains show differences in levels and regulation of enzymatic activity (Leathers et al., 1988; Manitchotpisit et al., 2009). Leathers (1986) showed that *A. pullulans* lineage produce xylanase constitutively at relatively high levels when grown on xylan, in addition to producing a significant amount of extracellular β-xylosidase. Typically, pigmented strains exhibited xylanase activity between 2.59 and 7.29 U mL^{-1}, while color-variant strains under the same conditions (optimum pH 4.0–4.5, optimum temperature 35–50°C) showed enzymatic activity ranging from 12.0 to 373.0 U mL^{-1} (Leathers, 1986). Manitchotpisit et al. (2009) based on the five loci sequencing of forty-five *A. pullulans* strains isolated from Thailand were classified into twelve clades. *A. pullulans* in some clades produced high xylanase activity (45.4 to 152.4 U mL $^{-1}$). Another strain of the species, *A. pullulans* CBS 135684, is known for producing xylanase with activity over a wide pH range (4.0–10.0) and at high temperatures (70–80°C) (Bankeeree et al., 2014). Recently, Yegin (2017) purified and analyzed xylanase produced by *A. pullulans* NRRL Y-2311-1 from wheat bran in the culture medium. The enzyme exhibited complete activity in the presence of 10% ethanol and was resistant to metals and reagents, such as Mg^{2+}, Zn^{2+}, Cu^{2+}, K^+, and EDTA, with an enzymatic activity of 848.4 U mL^{-1}.

Regarding β-xylosidase, studies show localization of the enzyme intracellularly or associated with the cell wall in *A. pullulans*. In the study by Christov et al. (1999), xylanase, purified from cultures of *A. pullulans*, was found secreted into the growth medium when grown on xylose and the enzyme β-xylosidase was cell-associated, as reported in other studies (Iembo et al., 2002). β-xylosidase in *A. pullulans* shows interesting properties in the activity of transxylosilation in the synthesis of new glycosidic bonds by transfer of a xylosyl residue from a donor molecule to a specific acceptor molecule, increasing its potential for the synthesis of new organic compounds. In addition to this property, β-xylosidase from *A. pullulans* also exhibits high stability in organic solvents and strong transfer activity (Ohta et al., 2010; Prasongsuk et al., 2018). In the study by Bankeeree et al. (2018), the cell-associated β-xylosidase produced by *A. pullulans* CBS 135684 is thermostable, being active at high temperatures (60–70°C) in a pH range of 5.0–9.0 and had a specific activity of 163.3 U mg^{-1}.

The use of xylanolytic enzymes produced by *A. pullulans* can be applied in the production of 2G ethanol, for example in SSF processes. Lara et al. (unpublished data) characterized the cell extract of *A. pullulans* UFMG-CM-Y518 and evaluated its potential in the enzymatic hydrolysis of wheat straw. *A. pullulans* extract showed increased activity at 40°C and pH ranging from 4.0 to 4.5 and optimal β-xylosidase activity at 80°C and pH 4.0 to 5.0, with high stability at a moderate temperature of 45°C. In SSF, using *Sp. Passalidarum* UFMG-CM-Y469 strain, *A. pullulans* UFMG-CM-Y518 extract efficiently degraded xylan to xylose with a higher yield (66%). This resulted in the production of 6.6 g L^{-1} of ethanol, achieving a yield of 0.2 g.g^{-1}. The UFMG-CM-Y518 extract improved the yield of ethanol from wheat straw by *Sp. Passalidarum* UFMG-CM-Y469 due to increased availability of xylose for fermentation.

During studies in yeasts associated with rotting wood and sugarcane bagasse from different regions of Brazil, Lara et al. (2014) detected xylanolytic activity in ascomycetous yeasts belonging to five species: *Candida intermedia, Candida tropicalis, Meyerozima guilliermondii, Sc. Shehatae,* and *Su. Smithiae.* Among basidiomycetous yeasts, four species had xylanolytic activity: *Naganishia diffluens, Kwoniella heveanensis, Papiliotrema laurentii,* and *Apiotrichum mycotoxinivorans.* The yeasts that reached the highest growth rate, cell density, and xylanase activity induced by xylan were the species *Pa. laurentii, Su. Smithiae,* and *Sc. Shehatae.* In cultures with media containing xylan as the only carbon source, *Pa. laurentii* showed the highest level of extracellular xylanolytic activity and *Su. Smithiae* achieved the highest level of β-1,4-xylosidase activity in the intracellular, extracellular and membrane-associated spaces. Under xylose induction, *Pa. laurentii* showed the highest specific activity of β-1,4-xylosidase at 50°C in the intracellular space among the yeasts tested. Morais et al. (2013b) described the species *Su. xylanicola,* isolated from rotting wood, which grows in a medium containing xylan as the sole carbon source and produces extracellular enzymes with xylanolytic activity. However, in the study by Lara et al. (2014), the species *Su. smithiae* showed higher xylanolytic activity than *Su. xylanicola* under the same experimental conditions. The species *Su. bahiana, Su. bonitensis, Su. carassensis, Su. ligni, Su. ayubii,* and *Su. valenteae* were isolated, identified, and their enzymatic activities were characterized by Sena et al. (2017). In addition, *Su. xylanicola* and *Su. lignohabitans* were isolated and both showed the highest xylanolytic activity, 0.29 U mL^{-1}. Among the strains of *Su. xylanicola* described by Morais et al. (2013) and Sena et al. (2017), the former had a lower value of enzymatic activity (0.11 U mL^{-1}), which may be justified by variations in metabolism between strains. Among the species of xylanolytic producers, *Su. smithiae* showed superior results in enzymatic activities induced by xylan (0.30 U mL^{-1} at 30°C and 0.50 U mL^{-1} at 50°C) (Lara et al. 2014). Handel et al. (2016) isolated *Su.*

mastotermitis and *Pa. odontotermitis*; however, the authors did not provide quantitative data on the enzymatic production by these yeasts.

Species of the genus *Scheffersomyces*, besides being great producers of ethanol, are also producers of xylanolytic enzymes. Özcan et al. (1991) purified and characterized the xylanase and β-xylosidase enzymes from *Sc. stipitis*, cultured in medium with xylan. While xylanase was secreted into the culture medium, significant amounts of β-xylosidase were found in the crude extract and bound to the cell wall, showing that xylanase production is dependent on xylan induction and is repressed by the presence of glucose and xylose in the culture medium. In the study by Lara et al. (2014), *Sc. shehatae* produced xylanase with a volumetric activity of 0.30 U mL^{-1} and 0.50 U mL^{-1} at temperatures of 30°C and 50°C, respectively.

CONCLUSION

A highly productive conversion of hemicellulose to ethanol is an important factor for the overall economy of the conversion process. Identification and collection of microorganisms capable of selectively converting D-xylose to ethanol at high rates have been the focus of extensive research in recent years. The study of these microorganisms contributes to a better understanding of the metabolism of pentoses, besides serving as a tool for introducing mutagenesis, protoplast fusion, breeding, genome shuffling, and adaptive evolution to generate artificial diversity. Beyond these strategies, metabolic engineering has led to the selection of superior industrial yeasts for natural xylose-fermenting strains and has been used for the development of recombinant microorganisms for strains of *S. cerevisiae*. Presently, *Sp. passalidarum* is considered the best xylose-fermenting yeast for the conversion of D-xylose to ethanol. For this reason, it is treated the best candidate for cultivation and/or as a source of genes for engineering yeasts strains, primarily *S. cerevisiae*, for 2G ethanol production. Xylose-fermenting yeasts are also sources of hydrolytic enzymes. Many species exhibit xylanase and β-xylosidase activities, and they have also been a donor of genes for genetic improvement of *S.*

cerevisiae. Therefore, yeasts with the ability to ferment xylose and/or produce enzymes that act on hemicellulose structure also have great potential for their application to the production of lignocellulosic ethanol.

ACKNOWLEDGMENTS

This work was funded by Conselho Nacional de Desenvolvimento Cientifico e Tecnológico (CNPq – Brazil, process numbers 407415/2013-1 and 0457499/2014-1), Fundação do Amparo a Pesquisa do Estado de Minas Gerais (FAPEMIG, process numbers APQ-01525-14, APQ-01477-13, APQ-02552-15).

REFERENCES

Agbogbo, F.K., Coward-Kelly, G., Torry-Smith, M. and Wenger, K.S. (2006). Fermentation of glucose/xylose mixtures using *Pichia stipitis*. *Process Biochemistry*, 41(11): 2333-2336.

Agbogbo, F.K. and Coward-Kelly, G. (2008). Cellulosic ethanol production using the naturally occurring xylose-fermenting yeast, *Pichia stipitis*. *Biotechnology Letters*, 30(9): 1515-1524.

Anwar, Z., Gulfraz, M. and Irshad, M., (2014). Agro-industrial lignocellulosic biomass a key to unlock the future bioenergy: a brief review. *Journal of Radiation Research and Applied Sciences*, 7(2): 63-173.

Arora, R., Sharma, N. K., Kumar, S., and Sani, R. K. (2019). Lignocellulosic Ethanol: Feedstocks and Bioprocessing. In *Bioethanol Production from Food Crops* (pp. 165-185). Academic Press.

Ayer, A., Gourlay, C.W. and Dawes, I.W. (2014). Cellular redox homeostasis, reactive oxygen species and replicative ageing in *Saccharomyces cerevisiae*. *FEMS Yeast Research*, 14(1): 60-72.

Azhar, S.H.M., Abdulla, R., Jambo, S.A., Marbawi, H., Gansau, J.A., Faik, A.A.M. and Rodrigues, K.F. (2017). Yeasts in sustainable bioethanol production: A review. *Biochemistry and Biophysics Reports*, *10:* 52-61.

Balat, M. (2011). Production of bioethanol from lignocellulosic materials via the biochemical pathway: a review. *Energy Conversion and Management*, *52*(2): 858-875.

Bankeeree, W., Akada, R., Lotrakul, P., Punnapayak, H. and Prasongsuk, S. (2018). Enzymatic hydrolysis of black liquor Xylan by a novel xylose-tolerant, thermostable β-Xylosidase from a tropical strain of *Aureobasidium pullulans* CBS 135684. *Applied Biochemistry and Biotechnology*, *184*(3): 919-934.

Bankeeree, W., Lotrakul, P., Prasongsuk, S., Chaiareekij, S., Eveleigh, D.E., Kim, S.W. and Punnapayak, H. (2014). Effect of polyols on thermostability of xylanase from a tropical isolate of *Aureobasidium pullulans* and its application in prebleaching of rice straw pulp. *SpringerPlus*, *3*(1): 1-11.

Bazoti, S.F., Golunski, S., Siqueira, D.P., Scapini, T., Barrilli, É.T., Mayer, D.A., Barros, K.O., Rosa, C.A., Stambuk, B.U., Alves Jr, S.L. and Valério, A. (2017). Second-generation ethanol from non-detoxified sugarcane hydrolysate by a rotting wood isolated yeast strain. *Bioresource Technology*, *244:* 582-587.

Belisa, B., Scheid, B., Gonçalves, D.L., Knychala, M.M., Matsushika, A., Bon, E.P. and Stambuk, B.U. (2015). Cloning novel sugar transporters from *Scheffersomyces (Pichia) stipitis* allowing d-xylose fermentation by recombinant *Saccharomyces cerevisiae*. *Biotechnology Letters*, *37*(10): 1973-1982.

Bozoudi, D. and Tsaltas, D. (2018). The multiple and versatile roles of *Aureobasidium pullulans* in the vitivinicultural sector. *Fermentation*, *4*(4): 85.

Brooks, R. E., Su, T. M., Brennan Jr, M. J., Frick, J. and Lynch, M. (1979). *Bioconversion of plant biomass to ethanol*. Final report, 15 December 1976-31 December 1978 (No. COO-4147-7). General Electric Co., Schenectady, NY (USA). Corporate Research and Development Dept.

Cadete, R.M. and Rosa, C.A. (2018). The yeasts of the genus Spathaspora: potential candidates for second-generation biofuel production. *Yeast*, 35(2): 191-199.

Cadete, R.M., Lopes, M.R. and Rosa, C.A., (2017). Yeasts associated with decomposing plant material and rotting wood. In *Yeasts in Natural Ecosystems: Diversity*: 265-292. Springer, Cham.

Cadete, R.M., Alejandro, M., Sandström, A.G., Ferreira, C., Gírio, F., Gorwa-Grauslund, M.F., Rosa, C.A. and Fonseca, C. (2016). Exploring xylose metabolism in Spathaspora species: XYL1. 2 from *Spathaspora passalidarum* as the key for efficient anaerobic xylose fermentation in metabolic engineered *Saccharomyces cerevisiae*. *Biotechnology for Biofuels*, 9(1): 1-14.

Cadete, R.M., Melo, M.A., Dussán, K.J., Rodrigues, R.C., Silva, S.S., Zilli, J.E., Vital, M.J., Gomes, F.C., Lachance, M.A. and Rosa, C.A., (2012). Diversity and physiological characterization of D-xylose-fermenting yeasts isolated from the Brazilian Amazonian Forest. *PLoS One*, 7(8): e43135.

Cadete, R.M., Santos, R.O., Melo, M.A., Mouro, A., Gonçalves, D.L., Stambuk, B.U., Gomes, F.C., Lachance, M.A. and Rosa, C.A. (2009). *Spathaspora arborariae* sp. nov., a D-xylose-fermenting yeast species isolated from rotting wood in Brazil. *FEMS Yeast Research*, 9(8): 1338-1342.

Canilha, L., Chandel, A.K., Suzane dos Santos Milessi, T., Antunes, F.A.F., Luiz da Costa Freitas, W., das Graças Almeida Felipe, M. and da Silva, S.S. (2012). Bioconversion of sugarcane biomass into ethanol: an overview about composition, pretreatment methods, detoxification of hydrolysates, enzymatic saccharification, and ethanol fermentation. *BioMed Research International*.

Cardona, C.A., Quintero, J.A. and Paz, I.C. (2010). Production of bioethanol from sugarcane bagasse: status and perspectives. *Bioresource Technology*, 101(13): 4754-4766.

Chandel, A. K., Garlapati, V. K., Singh, A. K., Antunes, F. A. F. and Silva, S. S. (2018). The Path Forward for Lignocellulose Biorefineries:

Bottlenecks, Solutions, and Perspective on Commercialization. *Bioresource Technology* 264: 370–81.

Chandel, A. K., Kapoor, R. K., Singh, A. and Kuhad, R. C. (2007). Detoxification of Sugarcane Bagasse Hydrolysate Improves Ethanol Production by *Candida Shehatae* NCIM 3501. *Bioresource Technology* 98 (10): 1947–50.

Christov, L. P., Myburgh. J., O'Neill, F. H., Tonder, T. A. and Prior, B. A. (1999). Modification of the Carbohydrate Composition of Sulfite Pulp by Purified and Characterized β-Xylanase and β-Xylosidase of *Aureobasidium Pullulans*. *Biotechnology Progress* 15 (2): 196–200.

Correia, K., Khusnutdinova, A., Li, P.Y., Joo, J.C., Brown, G., Yakunin, A.F. and Mahadevan, R. (2018). Flux balance analysis predicts NADP phosphatase and NADH kinase are critical to balancing redox during xylose fermentation in *Scheffersomyces stipitis*. *BioRxiv*, p.390401.

Cunha-Pereira, F., Hickert, L. R., Sehnem, N. T., Souza-Cruz, P. B., Rosa, C. A. and Ayub, M. A. Z. (2011). Conversion of Sugars Present in Rice Hull Hydrolysates into Ethanol by *Spathaspora arborariae*, *Saccharomyces cerevisiae*, and Their Co-Fermentations. *Bioresource Technology,* 102 (5): 4218–25.

Cunha, J. T., Soares, P. O, Thevelein, A. R. J. M. and Domingues, L. (2019). Xylose Fermentation Efficiency of Industrial *Saccharomyces cerevisiae* Yeast with Separate or Combined Xylose Reductase/Xylitol Dehydrogenase and Xylose Isomerase Pathways. *Biotechnology for Biofuels,* 12 (1): 1–14.

De Souza, R.D.F.R., Dutra, E.D., Leite, F.C.B., Cadete, R.M., Rosa, C.A., Stambuk, B.U., Stamford, T.L.M. and de Morais, M.A. (2018). Production of ethanol fuel from enzyme-treated sugarcane bagasse hydrolysate using D-xylose-fermenting wild yeast isolated from Brazilian biomes. *3 Biotech*, *8*(7): 312.

Delgenes, J. P., Moletta R. and Navarro, J. M. (1986). "The Effect of Aeration on D-Xylose Fermentation by *Pachysolen tannophilus*, *Pichia stipitis*, *Kluyveromyces marxianus* and *Candida shehatae*." *Biotechnology Letters,* 8 (12): 897–900.

Diderich, J. A., Schepper, M., Hoek, P. V., Luttik, M. A. H., Dijken, J. P. V., Pronk, J. T., Klaassen, P., Boelens, H.F., de Mattos, M.J.T., van Dam, K. and Kruckeberg, A.L. (1999). "Glucose Uptake Kinetics and Transcription of HXT Genes in Chemostat Cultures of *Saccharomyces cerevisiae*." *Journal of Biological Chemistry* 274 (22): 15350–59.

Ditzelmüller, G., Kubicek-Pranz, E.M., Röhr, M. and Kubicek, C.P. (1985). NADPH-specific and NADH-specific xylose reduction is catalyzed by two separate enzymes in *Pachysolen tannophilus*. *Applied Microbiology and Biotechnology*, 22(4): 297-299.

Ditzelmüller, G., Kubicek, C. P., Wöhrer, W., and Röhr, M. (1984). Xylose metabolism in *Pachysolen tannophilus*: purification and properties of xylose reductase. *Canadian journal of microbiology*, 30(11), 1330-1336.

Dumon, C., Song, L., Bozonnet, S., Fauré, R., and O'Donohue, M. J. (2012). "Progress and Future Prospects for Pentose-Specific Biocatalysts in Biorefining." *Process Biochemistry* 47 (3): 346–57.

Du Preez, J. C., Van Driessel, B., and Prior, B. A. (1989). "Ethanol Tolerance of *Pichia stipitis* and *Candida shehatae* Strains in Fed-Batch Cultures at Controlled Low Dissolved Oxygen Levels." *Applied Microbiology and Biotechnology* 30 (1): 53–58.

Du Preez, J. C., Bosch, M., and Prior, B. A. (1986)."Xylose Fermentation by *Candida shehatae* and *Pichia stipitis*: Effects of PH, Temperature and Substrate Concentration." *Enzyme and Microbial Technology* 8 (6): 360–64.

Du Preez, J. C., and Van der Walt, J. P. (1983). Fermentation of d-xylose to ethanol by a strain of *Candida shehatae*. *Biotechnology Letters* 362 (5): 357–62.

Dos Santos, L.V., de Barros Grassi, M.C., Gallardo, J.C.M., Pirolla, R.A.S., Calderón, L.L., de Carvalho-Netto, O.V., Parreiras, L.S., Camargo, E.L.O., Drezza, A.L., Missawa, S.K. and Teixeira, G.S. (2016). Second-Generation Ethanol: The Need Is Becoming a Reality. *Industrial Biotechnology* 12 (1): 40–57.

Farwick, A., Bruder, S., Schadeweg, V., Oreb, M., and Boles, E. (2014). Engineering of yeast hexose transporters to transport D-xylose without inhibition by D-glucose. *Proceedings of the National Academy of Sciences*, *111*(14), 5159-5164.

Gírio, F.M., Fonseca, C., Carvalheiro, F., Duarte, L.C., Marques, S. and Bogel-Łukasik, R. (2010). Hemicelluloses for Fuel Ethanol: A Review. *Bioresource Technology* 101 (13): 4775–4800.

Goshima, T., Tsuji, M., Inoue, H., Yano, S., Hoshino, T., and Matsushika, A. (2013). Bioethanol Production from Lignocellulosic Biomass by a Novel *Kluyveromyces marxianus* Strain. *Bioscience, Biotechnology, and Biochemistry* 77 (7): 1505–10.

Grootjen, D. R. J., Van der Lans, R. G. J. M., and Luyben, K. C. A. (1990). Effects of the Aeration Rate on the Fermentation of Glucose and Xylose by *Pichia stipitis* CBS 5773. *Enzyme and Microbial Technology* 12 (1): 20–23.

Gutiérrez-Rivera, B., Waliszewski-Kubiak, K., Carvajal-Zarrabal, O., and Aguilar-Uscanga, M. G. (2012). Conversion Efficiency of Glucose/Xylose Mixtures for Ethanol Production Using *Saccharomyces cerevisiae* ITV01 and *Pichia stipitis* NRRL Y-7124. *Journal of Chemical Technology and Biotechnology* 87 (2): 263–70.

Hahn-Hägerdal, B., Karhumaa, K., Fonseca, C., Spencer-Martins, I. and Gorwa-Grauslund, M.F. (2007). Towards industrial pentose-fermenting yeast strains. *Applied Microbiology and Biotechnology*, *74*(5): 937-953.

Hahn-Hägerdal, B. and Pamment, N. (2004). Microbial pentose metabolism. In *Proceedings of the Twenty-Fifth Symposium on Biotechnology for Fuels and Chemicals Held May 4–7, 2003, in Breckenridge, CO* (pp. 1207-1209). Humana Press, Totowa, NJ.

Hamacher, T., Becker, J., Gárdonyi, M., Hahn-Hägerdal, B. and Boles, E. (2002). Characterization of the xylose-transporting properties of yeast hexose transporters and their influence on xylose utilization. *Microbiology*, *148*(9): 2783-2788.

Handel, S., Wang, T., Yurkov, A.M. and König, H. (2016). *Sugiyamaella mastotermitis* sp. nov. and *Papiliotrema odontotermitis* fa, sp. nov. from the gut of the termites *Mastotermes darwiniensis* and

Odontotermes obesus. *International Journal of Systematic and Evolutionary Microbiology*, 66(11): 4600-4608

Hickert, L.R., de Souza-Cruz, P.B., Rosa, C.A. and Ayub, M.A.Z., (2013). Simultaneous saccharification and co-fermentation of un-detoxified rice hull hydrolysate by *Saccharomyces cerevisiae* ICV D254 and *Spathaspora arborariae* NRRL Y-48658 for the production of ethanol and xylitol. *Bioresource Technology*, *143:* 112-116.

Hossain, S.A., Švec, D., Mrša, V. and Teparic, R. (2018). Overview of catalytic properties of fungal xylose reductases and molecular engineering approaches for improved xylose utilisation in yeast. *Applied Food Biotechnology*, 5(2): 47-58.

Hou, X. (2012). Anaerobic xylose fermentation by *Spathaspora passalidarum*. *Applied Microbiology and Biotechnology*, 94(1): 205-214.

Iembo, T., Da Silva, R., Pagnocca, F.C. and Gomes, E. (2002). Production, Characterization, and Properties of β-Glucosidase and β-Xylosidase from a Strain of *Aureobasidium* sp. *Applied biochemistry and microbiology*, 38(6): 549-552.

Juturu, V. and Wu, J.C. (2013). Insight into microbial hemicellulases other than xylanases: a review. *Journal of Chemical Technology and Biotechnology*, 88(3): 353-363.

Karhumaa, K., Sanchez, R. G., Hahn-Hägerdal, B., & Gorwa-Grauslund, M. F. (2007). Comparison of the xylose reductase-xylitol dehydrogenase and the xylose isomerase pathways for xylose fermentation by recombinant *Saccharomyces cerevisiae*. *Microbial cell factories*, 6(1), 5.

Kilian, S.G. and Van Uden, N. (1988). Transport of xylose and glucose in the xylose-fermenting yeast *Pichia stipitis*. *Applied Microbiology and Biotechnology*, 27(5-6): 545-548.

Kim, H., Lee, H.S., Park, H., Lee, D.H., Boles, E., Chung, D. and Park, Y.C. (2017). Enhanced production of xylitol from xylose by expression of *Bacillus subtilis* arabinose: H+ symporter and *Scheffersomyces stipitis* xylose reductase in recombinant

Saccharomyces cerevisiae. Enzyme and Microbial Technology, 107: 7-14.

Kim, S.R., Ha, S.J., Kong, I.I. and Jin, Y.S. (2012). High expression of XYL2 coding for xylitol dehydrogenase is necessary for efficient xylose fermentation by engineered *Saccharomyces cerevisiae. Metabolic Engineering, 14*(4): 336-343.

Kirikyali, N. and Connerton, I.F. (2015). Xylan degrading enzymes from fungal sources. *Journal of Proteomics and Enzymology, 4*(1): 118.

Knoshaug, E.P., Vidgren, V., Magalhães, F., Jarvis, E.E., Franden, M.A., Zhang, M. and Singh, A. (2015). Novel transporters from *Kluyveromyces marxianus* and *Pichia guilliermondii* expressed in *Saccharomyces cerevisiae* enable growth on L-arabinose and D-xylose. *Yeast, 32*(10): 615-628.

Kumar, V. and Shukla, P. (2016). Functional Aspects of Xylanases toward Industrial Applications. In *Frontier discoveries and innovations in interdisciplinary microbiology* 157-165. Springer, New Delhi.

Kurtzman, C. P. and Suzuki, M. (2010). Phylogenetic Analysis of Ascomycete Yeasts That Form Coenzyme Q-9 and the Proposal of the New Genera Babjeviella, Meyerozyma, Millerozyma, Priceomyces, and Scheffersomyces. *Mycoscience* 51 (1): 2–14.

Kurtzman, C.P. (1983). Biology and physiology of the D-xylose fermenting yeast *Pachysolen tannophilus*. In *Pentoses and Lignin* (pp. 73-83). Springer, Berlin, Heidelberg.

Kwak, S., Jo, J.H., Yun, E.J., Jin, Y.S. and Seo, J.H. (2018). Production of biofuels and chemicals from xylose using native and engineered yeast strains. *Biotechnology Advances.* 37 (2): 271–83.

Kwak, S. and Jin, Y.S. (2017). Production of fuels and chemicals from xylose by engineered *Saccharomyces cerevisiae*: a review and perspective. *Microbial Cell Factories, 16*(1): 82.

Kwolek-Mirek, M., Maslanka, R. and Molon, M. (2019). Disorders in NADPH generation via pentose phosphate pathway influence the reproductive potential of the *Saccharomyces cerevisiae* yeast due to changes in redox status. *Journal of Cellular Biochemistry, 120*(5): 8521-8533.

Lara, C. (no prelo). Use of *Aureobasidium pullulans* xylanase for simultaneous saccharification and fermentation in second-generation bioethanol production.

Lara, C.A., Santos, R.O., Cadete, R.M., Ferreira, C., Marques, S., Gírio, F., Oliveira, E.S., Rosa, C.A. and Fonseca, C. (2014). Identification and characterisation of xylanolytic yeasts isolated from decaying wood and sugarcane bagasse in Brazil. *Antonie van Leeuwenhoek, 105*(6): 1107-1119.

Leandro, M.J., Fonseca, C. and Gonçalves, P. (2009). Hexose and pentose transport in ascomycetous yeasts: an overview. *FEMS Yeast Research, 9*(4): 511-525.

Leathers, T.D., Nofsinger, G.W., Kurtzman, C.P. and Bothast, R.J. (1988). Pullulan production by color variant strains of *Aureobasidium pullulans*. *Journal of Industrial Microbiology, 3*(4): 231-239.

Leathers, T. D. (1986). Color Variants of *Aureobasidium pullulans* Overproduce Xylanase with Extremely High Specific Activity. *Applied and Environmental Microbiology* 52 (5): 1026–30.

Li, X., Chen, Y. and Nielsen, J. (2019). Harnessing xylose pathways for biofuels production. *Current Opinion in Biotechnology, 57:* 56-65.

Limayem, A. and Ricke, S.C. (2012). Lignocellulosic biomass for bioethanol production: current perspectives, potential issues and future prospects. *Progress in Energy and Combustion Science, 38*(4): 449-467.

Liu, J.J., Zhang, G.C., Kwak, S., Oh, E.J., Yun, E.J., Chomvong, K., Cate, J.H. and Jin, Y.S., (2019). Overcoming the thermodynamic equilibrium of an isomerization reaction through oxidoreductive reactions for biotransformation. *Nature Communications*, 10(1), 1356.

Lopes, M.R., Batista, T.M., Franco, G.R., Ribeiro, L.R., Santos, A.R., Furtado, C., Moreira, R.G., Goes-Neto, A., Vital, M.J., Rosa, L.H. and Lachance, M.A. (2018). *Scheffersomyces stambukii* fa, sp. nov., a d-xylose-fermenting species isolated from rotting wood. *International Journal of Systematic and Evolutionary Microbiology.* 68: 2306-2312.

Lopes, M. R., Morais, C. G., Kominek, J., Cadete, R. M., Soares, M. A., Uetanabaro, A. P. T., Lachance, M., Hittinger, C. T., and Rosa, C. A.

(2016). Genomic analysis and D-xylose fermentation of three novel Spathaspora species: *Spathaspora girioi* sp. nov., *Spathaspora hagerdaliae* fa, sp. nov. and *Spathaspora gorwiae* fa, sp. nov. *FEMS yeast research*, *16*(4), fow044.

Manitchotpisit, P., Leathers, T.D., Peterson, S.W., Kurtzman, C.P., Li, X.L., Eveleigh, D.E., Lotrakul, P., Prasongsuk, S., Dunlap, C.A., Vermillion, K.E. and Punnapayak, H., (2009). Multilocus Phylogenetic Analyses, Pullulan Production and Xylanase Activity of Tropical Isolates of *Aureobasidium pullulans*. *Mycological Research* 113 (10): 1107–1120

Margaritis, A., and Bajpai P. (1982). Direct Fermentation of D-xylose to Ethanol by *Kluyveromyces marxianus* Strains. *Appl. Environ. Microbiol.*, 44:1039-1041.

Martins, D. A. B., do Prado, H. F. A., Leite, R. S. R., Ferreira, H., de Souza Moretti, M. M., da Silva, R. and Gomes, E. (2011). Agroindustrial wastes as substrates for microbial enzymes production and source of sugar for bioethanol production. *INTECH* Open Access Publisher.

Miskovic, L., Alff-Tuomala, S., Soh, K.C., Barth, D., Salusjärvi, L., Pitkänen, J.P., Ruohonen, L., Penttilä, M. and Hatzimanikatis, V. (2017). A design–build–test cycle using modeling and experiments reveals interdependencies between upper glycolysis and xylose uptake in recombinant *S. cerevisiae* and improves predictive capabilities of large-scale kinetic models. *Biotechnology for Biofuels*, *10*(1): 1-19.

Moon, J., Liu, Z.L., Ma, M. and Slininger, P.J. (2013). New genotypes of industrial yeast *Saccharomyces cerevisiae* engineered with YXI and heterologous xylose transporters improve xylose utilization and ethanol production. *Biocatalysis and Agricultural Biotechnology*, *2*(3): 247-254.

Morais, C.G., Batista, T.M., Kominek, J., Borelli, B.M., Furtado, C., Moreira, R.G., Franco, G.R., Rosa, L.H., Fonseca, C., Hittinger, C.T. and Lachance, M.A. (2017). *Spathaspora boniae* sp. nov., a D-xylose-fermenting species in the *Candida albicans*/Lodderomyces clade.

International Journal of Systematic and Evolutionary Microbiology, 67(10): 3798-3805.

Morais, C.G., Cadete, R.M., Uetanabaro, A.P.T., Rosa, L.H., Lachance, M.A. and Rosa, C.A. (2013). D-xylose-fermenting and xylanase-producing yeast species from rotting wood of two Atlantic Rainforest habitats in Brazil. *Fungal Genetics and Biology*, 60: 19-28.

Morais, C.G., Lara, C.A., Marques, S., Fonseca, C., Lachance, M.A. and Rosa, C.A. (2013b). *Sugiyamaella xylanicola* sp. nov., a xylan-degrading yeast species isolated from rotting wood. *International Journal of Systematic and Evolutionary Microbiology*, 63(6): 2356-2360.

Moreira, L.R.S. (2016). Insights into the mechanism of enzymatic hydrolysis of xylan. *Applied Microbiology and Biotechnology*, 100(12): 5205-5214.

Mosier, N., Wyman, C., Dale, B., Elander, R., Lee, Y.Y., Holtzapple, M. and Ladisch, M. (2005). Features of promising technologies for pretreatment of lignocellulosic biomass. *Bioresource Technology*, 96(6): 673-686.

Mussatto, S.I., Machado, E.M., Carneiro, L.M. and Teixeira, J.A. (2012). Sugars metabolism and ethanol production by different yeast strains from coffee industry wastes hydrolysates. *Applied Energy*, 92: 763-768.

Nguyen, N.H., Suh, S.O., Marshall, C.J. and Blackwell, M. (2006). Morphological and ecological similarities: wood-boring beetles associated with novel xylose-fermenting yeasts, *Spathaspora passalidarum* gen. sp. nov. and *Candida jeffriesii* sp. nov. *Mycological Research*, 110(10): 1232-1241

Niphadkar, S., Bagade, P. and Ahmed, S. (2017). Bioethanol production: insight into past, present and future perspectives. *Biofuels*, 9(2): 229-238.

Ohta, K., Fujimoto, H., Fujii, S. and Wakiyama, M. (2010). Cell-associated β-xylosidase from *Aureobasidium pullulans* ATCC 20524: Purification, properties, and characterization of the encoding gene. *Journal of Bioscience and Bioengineering*, 110(2):152-157.

Ota, M., Sakuragi, H., Morisaka, H., Kuroda, K., Miyake, H., Tamaru, Y. and Ueda, M. (2013). Display of *Clostridium cellulovorans* xylose isomerase on the cell surface of *Saccharomyces cerevisiae* and its direct application to xylose fermentation. *Biotechnology Progress*, 29(2): 346-351.

Özcan, S., Kötter, P. and Ciciary, M. (1991). Xylan-Hydrolysing Enzymes of the Yeast *Pichia stipitis*. *Applied Microbiology and Biotechnology*, 36: 190-195.

Panchal, C. J., Bast, L., Russell, I. and Stewart, G. G. (1988). Repression of xylose utilization by glucose in xylose-fermenting yeasts. *Canadian journal of Microbiology*, 34: 1316-1320.

Passoth, V. Conventional and non-conventional yeasts for the production of biofuels. *Yeast Diversity in Human Welfare*. Springer, 8: 385-416.

Prajapati, A. S., Panchal, K. J., Pawar, V. A., Noronha, M. J., Patel, D. H. and Subramanian, R. B. (2018). Review on cellulase and xylanase engineering for biofuel production. *Industrial Biotechnology*, 14(1): 38-44.

Prasongsuk, S., Lotrakul, P., Ali, I., Bankeeree, W., and Punnapayak, H. (2018). The current status of *Aureobasidium pullulans* in biotechnology. *Folia Microbiologica*, 63(2): 129-140.

Rech, F. R., Fontana, R. C., Rosa, C. A., Camassola, M., Ayub, M. A. Z. and Dillon, A. J. (2019). Fermentation of hexoses and pentoses from sugarcane bagasse hydrolysates into ethanol by *Spathaspora hagerdaliae*. *Bioprocess and Biosystems Engineering*, 42(1): 83-92

Ren, Y., Chen, L., Niu, Q. and Hui, F. (2014). Description of *Scheffersomyces henanensis* sp. nov., a new d-xylose-fermenting yeast species isolated from rotten wood. *PloS One*, 9 (3): e92315

Rocha, M. V. P., Rodrigues, T. H. S., Melo, V. M., Gonçalves, L. R., and de Macedo, G. R. (2011). Cashew apple bagasse as a source of sugars for ethanol production by *Kluyveromyces marxianus* CE025. *Journal of Industrial Microbiology and Biotechnology*, 38(8): 1099-1107.

Rodrussamee, N., Sattayawat, P. and Yamada, M. (2018). Highly efficient conversion of xylose to ethanol without glucose repression by newly isolated thermotolerant *Spathaspora passalidarum* CMUWF1-2. *BMC Microbiology*, *18*(1): 73.

Schneider, H., Wang, P.Y., Chan, Y.K. and Maleszka, R. (1981). Conversion of D-xylose into ethanol by the yeast *Pachysolen tannophilus*. *Biotechnology Letters*, *3*(2): 89-92.

Scordia, D., Cosentino, S.L., Lee, J.W. and Jeffries, T.W. (2012). Bioconversion of giant reed (*Arundo donax* L.) hemicellulose hydrolysate to ethanol by *Scheffersomyces stipitis* CBS6054. B*iomass and Bioenergy*, *39:* 296-305.

Sena, L.M., Morais, C.G., Lopes, M.R., Santos, R.O., Uetanabaro, A.P., Morais, P.B., Vital, M.J., de Morais, M.A., Lachance, M.A. and Rosa, C.A. (2017). D-Xylose fermentation, xylitol production and xylanase activities by seven new species of *Sugiyamaella*. *Antonie van Leeuwenhoek*, *110*(1): 53-6

Senthilkumar, V. and Gunasekaran, P. (2005). Bioethanol Production from Cellulosic Substrates: Engineered Bacteria and Process Integration Challenges. *Journal of Scientific and Industrial Research* 64 (11): 845–53.

Sharma, H.K., Xu, C. and Qin, W. (2019). Biological pretreatment of lignocellulosic biomass for biofuels and bioproducts: an overview. *Waste and Biomass Valorization*, *10*(2): 235-251.

Sharma, N.K., Behera, S., Arora, R., Kumar, S. and Sani, R.K. (2018). Xylose transport in yeast for lignocellulosic ethanol production: current status. *Journal of Bioscience and Bioengineering*, *125*(3): 259-267.

Singh, S., Sidhu, G.K., Kumar, V., Dhanjal, D.S., Datta, S. and Singh, J. (2019). Fungal Xylanases: Sources, Types, and Biotechnological Applications. In *Recent Advancement in White Biotechnology Through Fungi* 405-428. Springer, Cham

Slininger, P.J., Bolen, P.L. and Kurtzman, C.P. (1987). *Pachysolen tannophilus*: properties and process considerations for ethanol production from D-xylose. *Enzyme and Microbial Technology*, *9*(1): 5-15.

Slininger, P.J., Bothast, R.J., Van Cauwenberge, J.E. and Kurtzman, C.P. (1982). Conversion of D-xylose to ethanol by the yeast *Pachysolen tannophilus*. *Biotechnology and Bioengineering*, *24*(2): 371-384.

Su, Y.K., Willis, L.B. and Jeffries, T.W. (2015). Effects of aeration on growth, ethanol and polyol accumulation by *Spathaspora passalidarum* NRRL Y-27907 and *Scheffersomyces stipitis* NRRL Y-7124. *Biotechnology and Bioengineering*, *112*(3): 457-469.

Sun, Y. and Cheng, J. (2002). Hydrolysis of lignocellulosic materials for ethanol production: a review. *Bioresource Technology*, *83*(1): 1-11.

Urbina, H. and Blackwell, M. (2012). Multilocus phylogenetic study of the Scheffersomyces yeast clade and characterization of the N-terminal region of xylose reductase gene. *PLoS One*, *7*(6): e39128.

Varize, C.S., Cadete, R.M., Lopes, L.D., Christofoleti-Furlan, R.M., Lachance, M.A., Rosa, C.A. and Basso, L.C. (2018). *Spathaspora piracicabensis* fa, sp. nov., a D-xylose-fermenting yeast species isolated from rotting wood in Brazil. *Antonie van Leeuwenhoek*, *111*(4): 525-531.

Verduyn, C., Jzn, J. F., van Dijken, J. P. and Scheffers, W. A. (1985). Multiple forms of xylose reductase in *Pachysolen tannophilus* CBS4044. *FEMS Microbiology Letters*, *30*(3): 313-317.

Wang, S., Sun, X. and Yuan, Q., (2018). Strategies for enhancing microbial tolerance to inhibitors for biofuel production: a review. *Bioresource Technology*, *258*: 302-309.

Wang, Y., Ren, Y.C., Zhang, Z.T., Ke, T. and Hui, F.L. (2016). *Spathaspora allomyrinae* sp. nov., a D-xylose-fermenting yeast species isolated from a scarabeid beetle *Allomyrina dichotoma*. *International Journal of Systematic and Evolutionary Microbiology*, *66*(5): 2008-2012.

Wei, H., Yingting, Y., Jingjing, G., Wenshi, Y., and Junhong, T. (2017). Lignocellulosic Biomass Valorization: Production of Ethanol. *Encyclopedia of Sustainable Technologies (3), 601–604*.

Weierstall, T., Hollenberg, C.P. and Boles, E. (1999). Cloning and characterization of three genes (SUT1–3) encoding glucose transporters of the yeast *Pichia stipitis*. *Molecular Microbiology*, *31*(3): 871-883.

Wohlbach, D.J., Kuo, A., Sato, T.K., Potts, K.M., Salamov, A.A., LaButti, K.M., Sun, H., Clum, A., Pangilinan, J.L., Lindquist, E.A. and Lucas, S. (2011). Comparative genomics of xylose-fermenting fungi for enhanced biofuel production. *Proceedings of the National Academy of Sciences*, *108*(32): 13212-13217.

Yegin, S. (2017). Single-step purification and characterization of an extreme halophilic, ethanol tolerant and acidophilic xylanase from *Aureobasidium pullulans* NRRL Y-2311-1 with application potential in the food industry. *Food Chemistry*, *221*: 67-75.

Young, E., Poucher, A., Comer, A., Bailey, A. and Alper, H. (2011). Functional survey for heterologous sugar transport proteins, using *Saccharomyces cerevisiae* as a host. *Appl. Environ. Microbiol.*, *77*(10): 3311-3319.

Zhang, Y., Oates, L. G., Serate, J., Xie, D., Pohlmann, E., Bukhman, Y. V. Karlen, S. D., Young, M. K., Higbee, A., Eilert, D. and Sanford, G. R. (2018). Diverse lignocellulosic feedstocks can achieve high field-scale ethanol yields while providing flexibility for the biorefinery and landscape-level environmental benefits. *GCB Bioenergy*, *10*(11): 825-840.

INDEX

A

acetic acid, 7, 8, 21, 68, 113, 131
acid, vii, viii, x, 2, 4, 7, 58, 61, 64, 65, 68, 69, 75, 82, 113, 114, 115, 119
active site, 66
active transport, 120
additives, 33, 53, 54, 74, 79
alcohol production, 111
alcohols, viii, ix, 2, 5, 16, 19, 31, 32, 33, 46, 48, 49
algae, 27, 90, 91
alkane, ix, 32, 34, 37, 38, 47, 68
alternative energy, 59
aromatic hydrocarbons, 58, 59
aromatic rings, 69
aromatics, 42, 71, 78
assessment, 22, 24, 25
atmospheric pressure, 45

B

bacteria, 75, 114, 115
beetles, xi, 110, 125, 146
benefits, vii, x, 33, 58, 59, 90, 91, 150
benign, x, 23, 58
benzene, vii, x, 32, 34, 35, 37, 38, 39, 40, 41, 45, 46, 48, 49, 50, 78
biocatalysts, 116, 131
biochemical processes, 112
biochemistry, 142
bioconversion, 113, 114
biodegradability, x, 58, 59, 66, 67, 69, 92
biodegradation, 67, 68, 69
biodiesel, v, vii, viii, x, 1, 2, 3, 4, 5, 6, 7, 8, 9, 10, 11, 12, 13, 15, 16, 17, 18, 19, 20, 21, 22, 27, 46, 53, 57, 58, 59, 60, 61, 62, 63, 64, 65, 66, 67, 68, 69, 70, 71, 72, 73, 74, 75, 76, 77, 78, 79, 80, 81, 82, 83, 84, 85, 86, 90, 91, 92, 93, 94, 95, 96, 97, 98, 99, 100, 101, 102, 103, 104, 105, 106, 107
bioenergy, 22, 94, 98, 136
biofuels, v, vii, ix, 1, 3, 22, 24, 31, 32, 33, 34, 46, 50, 51, 52, 54, 75, 87, 89, 90, 91, 94, 98, 111, 138, 139, 143, 144, 145, 146, 147, 148
biological activity, 67
biological processes, 113

biomass, 5, 27, 52, 53, 86, 110, 112, 114, 115, 116, 136, 137, 138, 144, 146, 148
biotechnology, 147
black liquor, 137
blends, viii, ix, 31, 32, 33, 72, 73, 74, 76, 77, 79, 80, 81, 83, 85
bonds, 112, 113, 131, 133
Brazil, 62, 91, 109, 111, 127, 134, 136, 138, 144, 146, 149
butanol, v, vii, ix, 24, 26, 31, 32, 33, 34, 35, 37, 40, 41, 42, 44, 45, 46, 47, 49, 54, 55, 76, 78, 84, 97, 106, 107
butyl ether, 47, 48
by-products, 90, 91, 123

C

candidates, ix, 3, 6, 32, 33, 138
carbon, ix, x, 3, 9, 32, 33, 57, 59, 60, 62, 67, 73, 74, 76, 77, 84, 85, 92, 112, 113, 115, 121, 123, 129, 134
carbon atoms, ix, 32
carbon dioxide, x, 3, 9, 33, 57, 62, 67, 85
carbon emissions, 3
carbon monoxide, 3, 33, 59, 74, 76
carbon nanotubes, 77
carbon neutral, 62
catalyst, viii, x, 2, 20, 21, 58, 63, 64, 65, 66, 74, 77, 81, 83, 92
catalytic properties, 142
cell metabolism, 116, 122
cell surface, 123, 147
cellulose, 112, 113, 116, 131
chemical, viii, 1, 3, 11, 14, 15, 16, 19, 22, 23, 24, 41, 62, 63, 67, 68, 69, 82, 90, 112, 113
chemical industry, 11
chromatography, 34, 67, 68
coenzyme, 121, 122, 123
combustion, ix, x, 10, 32, 33, 46, 53, 58, 77, 79, 80, 81, 82, 85

commercial, 60, 65, 86, 93
complex carbohydrates, 114
composition, ix, 3, 32, 33, 35, 36, 37, 38, 39, 40, 68, 69, 75, 112, 138
compounds, viii, ix, 31, 32, 33, 35, 36, 42, 46, 48, 65, 66, 67, 69, 85, 90, 112, 115
computational fluid dynamics, 26
consumption, ix, 2, 32, 59, 62, 63, 86, 117, 120, 123, 127, 128, 129
contaminated soil, 68
cooking, 4, 7, 21, 53, 61, 64, 74, 75, 76, 77, 78, 79, 90, 92, 93
co-solvents, 9, 18
cost, 17, 19, 22, 23, 63, 65, 92, 110, 111, 112, 115
cost minimization, 22
cultivation, 86, 91, 135
culture, 112, 120, 127, 133, 135
culture medium, 127, 133, 135

D

degradation, vii, xi, 7, 67, 68, 69, 91, 110, 112, 113, 114
degradation process, 68
degradation rate, 69
detoxification, 118, 119, 138
developing countries, 63
dibutyl ether, vii, ix, 32, 34, 46, 47, 52, 53, 55
diesel engines, viii, 1, 3, 62, 75
diesel fuel, vii, ix, x, 3, 32, 33, 58, 59, 61, 69, 70, 72, 73, 74, 75, 76, 77, 78, 79, 80, 82, 84
diffusion, 66, 117, 126, 129
dispersion, 37, 39, 41
distillation, 10, 20, 22, 26, 27, 28, 86
distilled water, 75
distribution, ix, 32, 33

Index

E

economic disadvantage, 112
economic growth, 92
economics, 18
ecosystem, 67
education, 60
effluent, 24, 26, 27, 114
emission, x, 33, 58, 62, 66, 71, 72, 73, 75, 76, 77, 78, 79, 80, 81, 82, 83, 84
employment, 63
encoding, 120, 146, 149
endothermic, 37, 38, 39, 40, 41, 45
energy, vii, ix, x, 1, 2, 4, 10, 11, 17, 18, 19, 22, 32, 33, 39, 45, 50, 51, 52, 57, 58, 59, 60, 61, 68, 85, 92, 93, 96, 98, 99, 100, 103, 106, 110, 112, 121
energy consumption, x, 19, 57, 58, 92
energy efficiency, 33, 50, 51
energy input, 11, 17, 85
energy integration, 10, 22
energy supply, 60, 111
engineering, 13, 50, 51, 52, 95, 102, 117, 120, 123, 131, 135, 142, 147
enthalpy, v, vii, ix, 31, 32, 34, 36, 37, 38, 39, 40, 41, 42, 43, 44, 45
environment, 4, 58, 60, 68, 97
environment effects, 58
environmental characteristics, 112
environmental impact, 3, 10, 11, 18, 33, 61, 86, 91
environmental issues, x, 57, 92
environmental protection, 67
enzymatic activity, 120, 128, 132, 134
equipment, viii, 2, 5, 10, 12, 14, 15, 16, 23, 29, 65
ester, 63, 66, 68, 69, 70, 74, 81, 82, 84
ethanol, ix, xi, 5, 9, 10, 20, 21, 24, 26, 32, 33, 48, 59, 60, 65, 72, 73, 74, 78, 79, 85, 91, 109, 110, 111, 112, 113, 115, 116, 120, 123, 124, 126, 127, 128, 129, 130, 131, 133, 135, 136, 137, 138, 139, 140, 142, 145, 146, 147, 148, 149, 150
experimental condition, 130, 134

F

fatty acids, viii, 1, 3, 5, 7, 8, 19, 66, 67, 69, 76, 77
feedstock, vii, x, 5, 27, 58, 63, 64, 66, 67, 68, 77, 91, 93, 110, 112
fermentation, xi, 24, 26, 86, 109, 112, 114, 115, 116, 117, 118, 119, 120, 121, 125, 126, 127, 128, 129, 130, 131, 134, 137, 138, 139, 142, 143, 144, 145, 147, 148
fermentation technology, 112
first generation, 86, 87, 90, 111
flexibility, 14, 150
food, 64, 85, 86, 90, 91, 150
food chain, 85, 90, 91
food industry, 150
food security, 86
formaldehyde, 72, 78
formation, viii, 2, 12, 59, 66, 76, 112, 120, 123

G

genes, 120, 123, 126, 128, 131, 135, 149
genus, 125, 127, 128, 135, 138
global demand, 86
global warming, x, 3, 58, 59, 61, 92
gluconeogenesis, 115, 121
glucose, xi, 109, 111, 112, 115, 117, 119, 120, 126, 128, 129, 130, 135, 136, 141, 142, 147, 149
glucosidases, 114, 132
glycerin, 20, 63, 65
glycerol, 3, 7, 12, 20, 21, 63, 120, 123
glycolysis, 115, 121, 128, 145
greenhouse gas emissions, ix, 32, 33, 45, 79, 91

H

growth, 58, 60, 68, 71, 72, 78, 79, 85, 90, 91, 133, 134, 143, 149

H

harmful effects, 61, 112
hazardous materials, 105
hazardous substance, 17
hazards, vii, viii, 2, 15, 17, 18
hemicellulose, vii, xi, 110, 113, 114, 116, 131, 135, 148
hemisphere, 61
heptane, vii, ix, 32, 34, 35, 37, 38, 39, 40, 41, 45, 46, 47, 48
homogeneous catalyst, 19, 66
homologous genes, 117
hydrocarbons, vii, ix, x, 32, 33, 34, 42, 45, 49, 58, 62, 68, 69, 92
hydrolysis, vii, viii, xi, 2, 5, 8, 110, 112, 114, 119, 132, 133, 137, 146
hydroxide, viii, 2, 63, 65, 113

I

incomplete combustion, 60, 62
industry, x, 2, 11, 22, 32, 34, 50, 51, 61, 146
infrastructure, ix, 32, 33
inherent safety, viii, 2, 4, 14, 15, 16, 23, 24
insects, xi, 110, 124, 130
integration, viii, 2, 10, 11, 18, 20, 22, 28, 120
isolation, 131, 132
isomerization, 121, 123, 144

K

kinetic constants, 121
kinetic model, 120, 145
kinetic parameters, 121
kinetics, 53, 117

L

lignocellulose, vii, xi, 109, 110, 112, 138
liquefied natural gas, x, 58, 59
low temperatures, 3, 113

M

materials, 53, 58, 66, 113, 114, 130, 132, 137, 149
media, 63, 123, 129, 134
metabolism, 115, 121, 122, 123, 125, 126, 128, 131, 134, 135, 138, 140, 141, 146
metal oxides, 65, 66
metal salts, 66
methanol, viii, ix, x, 2, 3, 5, 6, 7, 8, 9, 10, 12, 19, 20, 21, 22, 32, 33, 58, 63, 72, 74, 113
methodology, 14, 15, 16, 24
methyl group, 38, 42
microemulsion, 62, 92
microorganism, xi, 109, 115, 116, 121
microorganisms, 68, 69, 112, 115, 120, 132, 135
mixing, ix, 32, 34, 35, 36, 39, 49, 82, 84
modifications, ix, 32, 34, 79, 123
molecules, 36, 37, 38, 39, 41, 48, 66, 85

N

non-catalytic production, v, 1, 2, 8

O

octane, vii, ix, 31, 32, 33, 34, 35, 37, 38, 39, 40, 41, 45, 47
octane number, ix, 31, 32, 33, 34
oil, 5, 7, 9, 18, 19, 21, 27, 45, 53, 58, 59, 61, 63, 66, 69, 71, 72, 73, 74, 76, 77, 78, 79, 80, 81, 82, 83, 85, 87, 88, 89, 90, 91, 92

Index

optimization, 13, 14, 20, 21, 25, 27, 28
organic compounds, 133
organic solvents, 48, 133
ox, viii, ix, x, 31, 32, 33, 34, 42, 46, 53, 84
oxidation, 76, 80, 121
oxide nanoparticles, 78
oxygen, x, 39, 58, 62, 67, 68, 71, 73, 76, 77, 80, 81, 82, 83, 122, 123, 124, 125, 126, 128

P

pathway, 115, 121, 123, 125, 128, 137, 143
petroleum, ix, x, 2, 32, 57, 59, 63, 68, 69, 70, 92
phenolic compounds, 112
phosphate, 66, 121, 123, 125, 128, 143
phosphorus, 91
plants, 11, 13, 16, 18, 22, 23, 59, 61, 65, 85, 92
polycyclic aromatic hydrocarbon, 68, 69
process intensification, viii, 2, 9, 10, 11, 17, 18, 22, 26, 27, 28
producers, 111, 128, 129, 132, 134, 135
production costs, 4, 63, 111
protection, 12, 15, 122
proteins, 117, 122, 131, 150
purification, viii, 2, 3, 7, 10, 24, 26, 114, 123, 140, 150

R

raw materials, 5, 14, 17, 33, 92, 111, 112
reactants, viii, 2, 4, 9, 11, 65
renewable energy, 3, 60, 61
renewable energy technologies, 60
requirement, 10, 59, 128
requirements, vii, viii, 2, 4, 10, 11, 91
resources, x, 58, 60, 61, 66, 92, 110
risk, viii, 2, 4, 11, 12, 13, 14, 15, 18, 22, 23, 33, 94

S

safety, viii, 2, 4, 11, 12, 14, 15, 16, 18, 23, 24, 26, 28, 29
saturated fat, 67, 69, 77
saturated fatty acids, 67, 69, 77
scientific papers, 25, 28
second generation, ix, 32, 34, 89, 90, 112
second-generation ethanol, vi, 109, 110, 137, 140
seed, 68, 73, 84, 85, 87, 88, 89, 90
showing, 38, 69, 78, 127, 135
silica, 66
simulation, ix, 5, 19, 27, 32, 34
social benefits, x, 58
sodium, viii, 2, 3, 63, 65, 113
sodium hydroxide, 3, 63
species, xi, 109, 116, 125, 126, 127, 128, 129, 130, 132, 134, 135, 136, 138, 144, 145, 146, 147, 148, 149
stress, 115, 116, 122
stress response, 122
strong interaction, 37, 39
structure, vii, xi, 16, 62, 67, 110, 136
substrate, xi, 66, 109, 123, 129
substrates, xi, 110, 145
sugarcane, 111, 113, 129, 131, 134, 137, 138, 139, 144, 147
sulfur, 59, 70, 71, 72, 113
sulfuric acid, viii, 2, 3, 64, 65, 113
supercritical processes, vii, viii, 2, 4, 7, 9, 11, 18
surface area, 113
surface tension, 77
sustainability, x, 16, 24, 58
sustainable development, 60
sustainable economic growth, 59
sustainable energy, v, 27, 58, 93, 96, 104, 105
synthesis, ix, 9, 19, 21, 23, 28, 32, 34, 63, 122, 133

Index

T

technology, 5, 20, 21, 22, 50, 63, 100, 101, 104, 112, 115
temperature, 4, 7, 9, 11, 12, 18, 20, 33, 35, 36, 39, 41, 42, 45, 47, 48, 53, 54, 60, 63, 66, 77, 80, 81, 113, 126, 132, 133
thermal degradation, 5
thermal energy, 10, 50, 51, 52
thermal stability, 6
thermal treatment, 113
thermodynamic equilibrium, 123, 144
thermodynamic properties, 33, 36, 49, 50, 53
thermodynamics, 50, 51, 52
toluene, vii, x, 32, 34, 35, 37, 38, 39, 40, 41, 45, 47, 48, 78
total energy, 10
toxic metals, 70
toxicity, x, 58, 70, 92
transesterification, viii, 1, 4, 5, 9, 19, 21, 62, 63, 65, 92
transformation, xi, 110, 115, 116
transport, vii, ix, x, 1, 3, 12, 15, 32, 34, 116, 117, 120, 121, 126, 129, 140, 144, 148, 150
transport processes, 116
transportation, 58, 61, 92, 111
treatment, viii, 1, 3, 7, 19, 112, 113, 114, 119
triglycerides, viii, 1, 3, 5, 7, 19, 21, 86

V

vegetable oil, viii, x, 1, 3, 4, 5, 19, 58, 59, 61, 62, 64, 65, 66, 67, 69, 70, 72, 90, 93

viscosity, 59, 62, 76, 77, 82

W

waste, 4, 7, 19, 21, 53, 63, 64, 67, 74, 75, 76, 77, 78, 79, 83, 91, 93
water, ix, 5, 6, 7, 10, 19, 32, 33, 60, 71, 81, 83, 84, 86, 90, 91, 132
wind power, 3, 60, 61
wind turbines, 61
windows, 60
wood, xi, 109, 124, 125, 127, 129, 131, 134, 137, 138, 144, 146, 147, 149
workers, 13
worldwide, 126

X

xylanolytic enzymes, vii, xi, 110, 131, 133, 135
xylose-fermenting yeasts, xi, 110, 116, 124, 125, 126, 135, 138, 146, 147

Y

yeast, xi, 75, 86, 109, 115, 116, 118, 121, 123, 124, 126, 128, 129, 130, 131, 132, 135, 136, 137, 138, 139, 140, 141, 142, 143, 145, 146, 147, 148, 149, 150
yield, vii, xi, 5, 9, 110, 115, 123, 124, 128, 129, 130, 133